W9-ATO-606

Facing Starvation

Facing Starvation

Norman Borlaug and the Fight Against Hunger

by

Lennard Bickel

READER'S DIGEST PRESS
Distributed by E. P. Dutton & Co., Inc., New York 1974

Library of Congress Cataloging in Publication Data

Bickel, Lennard.
Facing starvation: Norman Borlaug and the fight
against hunger.

1. Borlaug, N. E.   I. Title.
HD9019.B66B5   1974   630'.92'4 [B]   74-1078
ISBN 0-88349-015-3

First Edition

10 9 8 7 6 5 4 3 2 1

To "Stak,"
Elvin C. Stakman,
paterfamilias to the
Hunger Fighters

Foreword

As I flew from New York to Mexico City for my final interviews with Norman Borlaug, I was struck by the paradox of his situation. Here was a man who had won the Nobel peace prize for his incredibly unselfish labors to provide food for hungry people around the world. He had been hailed as the father of the Green Revolution for having created new strains of high-yield disease-resistant wheat, and for his missionary efforts to bring the fruit of his research to the agricultural backwaters of the world. And he probably knew better than anyone else just how perilous our world food problem has become.

Yet Borlaug's voice has been a lonely one in our civilized world of plenty.

He has always labored in the fields. During the past twenty-six years his strong hands have been grimed by dirt and burned by sun. His blue eyes have the look of a man who has seen everything. He knows sweat; he knows bone-crushing fatigue. He understands the relationship of man to earth, and the bond between earth and the plants that feed mankind. He knows the reality of starvation. He does not dwell in an ivory tower. He does not spend his time writing articles for learned journals or debating figures on population growth, harvest estimates, or subsidies. He is dedicated above all else to the concept that it takes common sense and hard work to produce the food needed to fill empty bellies. Perhaps it is because of all these things that the people who live in cities do not, or cannot, hear his message.

Having followed Borlaug around the world for two years, I have witnessed the glories of his accomplishments and the shattering suffering that he has pledged himself to eradicate. Thus I have no doubts left about the correctness of Borlaug's thoughts or actions. His knowledge should be understood throughout the world. The tragedy is that understanding may come too late.

On that flight to Mexico last year, I read the latest clippings my editors had saved for my trip. One national weekly magazine quoted America's farm experts as saying that America's need for wheat could be satisfied by holding 400 million bushels of wheat in storage. But, the articles went on to say—all too blandly, I felt—these same experts predicted that by the end of 1974 America would have only half the reserve it needed. My concern increased on reading another clip from an afternoon New York newspaper. It was a story reporting that the world was on "the brink of mass hunger." Yet this item was only 14 lines long and was placed on the ninety-second page of a ninety-four-page newspaper. Next was an item in a morning national newspaper pointing out that America's 1974 wheat crop might not be the bumper harvest Washington hoped for. The reasons were many, but three were paramount. The first was the fact that fall rains had washed out nearly 750,000 acres of winter wheat. The second was that farmers either had no land left to put under the plow or were afraid to grow more wheat because it might create another dustbowl disaster as had occurred in the 1930s. Third, there was going to be a shortage of fertilizer. Most worrisome of all was another brief item—only six lines long—reporting that the White House had awakened to the "peril of future food crises" but since the 1973 harvest had been so good, it would be difficult to capture any "presidential attention" for a long-term plan until next year.

Finished with my reading, and more disturbed than I cared to admit, I was eager for my plane to land so I could question Borlaug about the significance of these stories. How incredible, I thought to myself, that I am going to talk to one of the few men in the world who really understands the gravity of the situation.

Norman Borlaug came to the Mexico City airport to meet me. One cuff of his khaki twill work pants was caught in the rear tab of his ankle-high boots. He was covered with dust as the result of having just finished harrowing a field for planting, but his grin was as wide and friendly as ever. What a contrast he was to the well-dressed international travelers at the airport.

In the days that followed I trudged behind him through plot after plot of experimental wheats and a new cereal grain he is working on, triticale. And I listened to him talk to the young men from around the world who he is training in the latest techniques of growing grain. More than twenty nations had sent students to learn his techniques: there were Turks, Pakistanis, Poles, Algerians, Russians, Bangladesh, Burmese, Israelis, Africans, and South Americans. While Borlaug showed them how to prepare the soils, analyze the proper requirements for fertilizer, and cross-breed wheat, he kept up a running commentary.

"I fear that we will have a worldwide disaster—possibly as soon as next year—before we comprehend the gravity of the situation," he said. "And it is unthinkable that mankind must face starvation before we learn that the people of every nation must be fed. How critical is our situation? There is less than a 1 percent margin of reserves left in the world food grain supply. I know this is difficult to believe, especially when the world is producing around 1.4 billion tons of grain every year. But it is true. We have less than a 1 percent margin for error before untold millions of people starve to death. The disaster that faces us can come in almost any form: droughts in Africa or Russia, too much rain in India and Southeast Asia, a plant disease such as rust infestation in America. Despite all our technology, we have no defense against bad weather or plant disease. The Bible tells us that Joseph saved Egypt from starvation by storing seven years' reserves of grain. But as of this moment the world has depleted nearly all the grain stocks in its food bins. In 1973 America took 25 million acres out of the 50 million acres of land we have left in reserve and put it into grain crops. You might say America has lots of land left in reserve. Well, no one really knows how good it is, or even how many acres of it will grow crops. And I hear the dreadful ticking of the biological clock. Five births occur every two seconds. That means over 70 million new mouths to feed every year. To put it another way, a new nation the size of the United States or Russia is now appearing on earth every three years. We are in a race against time to keep from starving."

Borlaug's voice is not bitter nor angry nor proselytizing, but calm and reasoning and marked by the Norwegian accent that can still be heard in the wheat and corn fields of his native Iowa. He is proud of the fact that the Green Revolution began in this Mexican valley, and that his wheats created the greatest increase of grain production the world has known.

How does one value the miracle Borlaug has accomplished? It is difficult. Let us use the benchmark year of 1965 when India produced only 12.5 million tons of her own wheat. In 1972, the Green Revolution created by Borlaug gave India 26.5 million tons of wheat, which made it the third largest producer of wheat in the world. Most of us would value the tonnage differential at the market price of more than two hundred dollars a ton. Not Borlaug. He prefers to look at it another way: each ton of wheat feeds five people for a whole year. Thus his wheat provided a year's nourishment for an additional 70 million people in India. There are numerous other countries in which he has achieved similarly spectacular successes.

Most of us would say that the Green Revolution represents a great victory. After all, providing nourishment for 70 million people in one nation alone is an incredible achievement. But again Borlaug would disagree. He sees it as little more than a rest period so that the world can take stock of the situation and prepare for the struggles yet to come. A man such as Borlaug knows the realities of famine: thousands of villages in hot, dusty lands where drought means doom, where the last blackened stalks emerging from the parched earth are devoured by hordes of ravenous rats; where dark skins are tight over protruding bones and children with swollen bellies and spindly legs look at visitors with pleading eyes; where red tints in black hair scream "protein deficiency"; where a handful of grain means life.

While many of us sit back and relax, Borlaug worries about the growing population in the world. He worries about the interlocking relationships that exist between man and nature, between one continent and another, and about the effects of price increases on people who cannot buy food. Borlaug looks at his visitor and now his blue eyes flash. "There can be no real and lasting comfort in any corner of the world when one talks about food," he says. "We dream of isolation—of being free of the effects and outcome of the spreading problem of hunger. Yet our isolation will be shattered by poverty and misery in immense areas of the earth.

"America, Australia, Canada, Europe—these places can

no longer think of themselves as sanctuaries divorced from the rest of mankind. Consider the new problems in America alone. Less than 5 percent of our population is engaged in producing food. Only one in eight members of Congress has any understanding of agricultural problems. The mass of urban dwellers that make up America today forget we can still have agricultural disasters. We, and by this I mean the farmers, deal with biological patterns that change and evolve. America lost almost its entire corn crop in 1971 when the Southern leaf blight came out of nowhere and destroyed close to 700 million bushels. We escaped that disaster by the skin of our teeth, only because our science was prepared for such a problem. Disaster can happen again. Not just once, but for several years in a row. We don't have any defense against that. Think what would happen to the American people. Then think what would happen to the rest of the world.

"The first agonies will be those of the poorer peoples. We would not be able to export wheat to other countries. Grain wouldn't be available at any price. Without grain, cattle can't be fed; bread can't be baked. People will die by the millions. It is terrifying to think about. We must think about it, however, and we must prepare ourselves for the battles yet to come."

My interviews finished, I took leave of Borlaug in a greenhouse in which he was growing his newest crosses of grain. And as my plane took off from Ciudad Obregón, I glanced down in the warm afternoon light at fields of brilliant green wheat growing on land that twenty years ago had been declared not worthy of farming. And I could not help but wonder if Borlaug's worry of disaster in 1974 would come true. Well, I thought, by the time I reach the last chapters I'll know the answer and I can write about it.

Meanwhile, I had the task of finishing this book about a man who has dedicated his life to fighting hunger, poverty, misery, ignorance, sloth, and selfishness. It is only fitting to his character and values that the battle he wages is such an epic encounter. I recalled the words of Walt Whitman, who wrote:

> Now I see the secret of making the best persons;
> It is to grow in open air,
> And to eat and sleep with the earth.

The wonders and beauties of nature exist for Borlaug as for few other men. I can never forget his explanation about why he hated paperwork and why he had resisted every attempt to get him to fulfill bureaucratic tasks. "I am poor at administrative work, impatient," he said. "When I start writing, my thoughts race ahead of my pencil and the whole writing process gets jammed up. But it is different with wheat. I can talk to the damned plant. And it talks back to me in a low voice that can't be heard in an office. It tells me if it's happy, healthy, if it has real character, or if it's just *asi, asi* (so, so). It speaks to me first through my eyes. It either looks healthy or it doesn't. And it smells healthy. A healthy stalk has a slick feeling; you can feel a pastule of rust before you can see it. Then you can cup your hand, feel the weight of the seed in the heads of the wheat, and you can really tell if it's producing. When wheat is ripening properly, when the wind is blowing across the field, you can hear the beards of the wheat rubbing together. They sound like the pine needles in a forest. It is a sweet whispering music that once you hear you never forget."

In shaping or reconstructing nature with the tools of his science, Borlaug uses the designs of nature wisely—for the perpetuation of the human race. He does so with knowledge, reason, compromise mixed with driving force, and a consuming flame. In doing this he has also paid a price. His has been a life of selflessness and hardship. He and his wife,

Margaret, have sacrificed much for what has been achieved, and they know that more sacrifices lie ahead of them.

The Nobel peace prize struck Borlaug like a typhoon. No other agronomist had been so honored and he had not expected it; nor had he wanted it. One of the last things he told me was: "I have always loved solitude, and my solitude has been under fire ever since I was awarded the peace prize. It brought more challenges. It challenged my sense of obligation to my fellow man. Margaret and I have wondered whether it was worth it. I often wish that I could recapture that beautiful solitude I knew when I was young and on the mountain—above the Salmon River in Idaho—but I shall never know such lovely peace again."

It is my hope that this biography of Norman Borlaug will reveal the complexity of battles he has fought, and the true greatness and character of one of the noblest men of our time.

Lennard Bickel
Cronulla NSW
Australia

Acknowledgments

The scope of Norman Borlaug's life and work is global; thus I am in debt to many people in different lands. It would be impossible to mention them all here, but to all whose names escape this list I extend my gratitude.

I am deeply beholden to Norman Borlaug and his wonderful wife, Margaret, for unfailing kindness, courtesy, availability. It is obvious that without their cooperation the work would never have been completed. Among the many others who made this work possible none is so crucial as the late Dr. John Falk, former chief of the plant industry division of Australia's CSIRO, who first introduced me to Norman Borlaug in 1968. A superb scientist, imbued with the fire that burns in hunger fighters, John Falk believed that man could meet future food needs with expanded science, and when he died in 1970 he left me with the resolve to write the Borlaug story.

The work has been aided by others with the same spirit. Paramount among them is Dr. J. George Harrar, who gave of his busy hours when he was President of the Rockefeller Foundation, and Dr. Glenn Anderson of Canada, who was deeply involved with the grain explosion in India and other lands and still found time to help. Dr. Keith Finlay, and

Dr. Robert Osler of CIMMYT, were constant stalwarts, and CIMMYT teems with others who contributed records and recollections: the Director-General, Haldore Hanson, whose name still resounds in Pakistan, the patient and kindly Dr. Ignacio Narváez, Dr. Evangelina Villegas and members of her nutritional staff, and, not least, the beloved and respected Dr. Edwin Wellhausen. Their companions in the field were of no less help—the incomparable Dr. Frank Zillinsky, Dr. John Lindt, Reyes Vega, and Drs. Arnoldo Amaya, Armando Campos, Tony Fischer, Sanjaya Rajaram, among other scientists.

I am also in debt to the late Dr. Joseph Rupert, to Dr. Orville Vogel, and most especially to the "Three Horsemen" who formed the Rockefeller commission to Mexico: Professors E. C. Stakman, Richard Bradfield and Paul C. Mangelsdorf. Their writings and recollections have been invaluable.

Grateful recognition is due also to Mr. Wayne Swegle, formerly managing editor of *Successful Farming,* for the help freely given from his studies in Pakistan for the Ford Foundation. In India, Mr. C. Subrimaniam and Dr. M. S. Swaminathan used their good offices to help as did the Australian High Commissioner, Sir Patrick Shaw. I am also indebted to the Indian government's information services.

In Mexico I am especially grateful to Teresa and Roberto Maurer, to Richard Spurlock for long hours of interview, and to José Cadena for local legend and lore.

The Rockefeller Foundation has been generous with its help on publications, and Dr. Lowell Hardin of the Ford Foundation gave welcomed assistance.

Gratefully, I recognize the superb skill with which my editors have molded a copious narrative into a story of manageable proportion; to Bruce Lee, Nancy Kelly and Jane Gunther I owe the most humble thanks for patience, solicitude, and tolerance, in the polishing of a rough stone.

Assistance in the collection and compilation of research material, in reading, writing letters, and manuscript preparation came unstintingly from Evangelina C. de Viesca, Dr. Borlaug's secretary; from my own dear wife, Pauline; and from Miss Geraldine O'Sullivan and Mrs. Olga Moore. Their aid was invaluable.

I am also grateful to my fellow Australians, Sir Otto Frankel, internationally known geneticist, and to Sir John Crawford for his expertise on India. But most of all, like Norman Borlaug, I could not conclude my appreciations without thanks to the millions of farmers across the world who made the Green Revolution a reality and to the army of hunger fighters in many lands who continue to seek hope for the future.

Lennard Bickel
January, 1974

Facing Starvation

One

It was a freezing January day. Bone-chilling winds swept southward, down across the Minnesota border, driving snow banks against the rural school of New Oregon Township Number Eight. The school was located far from any habitation—five miles from the village of Saude and fifteen miles from the railhead town of Cresco. The site had been pinpointed on a map of northeast Iowa by some unknown planner following the arbitrary stipulation that country schools were to be a certain number of miles from one another. Lumber for the one-room structure had been freighted to the designated site, nailed together, and painted a color long since faded, and the building had been left to fulfill its functions with a rhythm that never seemed to change.

Teachers came one year and left the next. In the hot days of spring and summer the windows were open. Now, in the winter of 1920, they were firmly shut, and the potbellied iron stove was red hot in its never-ending battle against the chill. As was customary, one teacher was responsible for all eight grades. Winter was her time to work hard with the older students. In spring, summer, and fall they would be needed to work the fields of their parents' farms. Some came to school during the winter months only, and would be eighteen by the time they reached eighth grade. The teacher would concentrate on the younger students another time.

It was only natural, then, that five-year-old Norman Ernest Borlaug sat daydreaming at his first-grade desk, watching the snowflakes skim past the windows in fascinating patterns. It had been like that all morning. Some snow was drifting in under the door, and it was still falling at lunchtime. The children ate their sandwiches at their desks; they were allowed to go outside to the toilet, but not to play. He saw the weathered shingle roof of the schoolhouse covered like an iced cake. Fragments of snow clung to the walls where the paint had peeled away, and in the eaves, near the outlet flue from the potbellied stove, the snow had melted and formed icicles ten inches long.

It stopped snowing after lunchtime; but the wind began to batter harder at the door. It grew colder in the classroom and the teacher threw more wood into the stove, until the flames roared in the iron flue. Still the children shivered. The room grew gloomier. The moisture from the children's breath froze into delicate fern patterns on the windows. The teacher lit oil lamps and continued with the lessons. After a time she went to the door and looked out anxiously at great blue-black clouds rising from the north, clouds with ominous, gray edges.

Long before the normal time to end school she called them out from behind their desks and told them to pull their scarves over their heads.

She spoke to the older children. "You will have to go home Indian file. The sky is very threatening and it will get a lot colder. Get them home, boys, as quick as you can." She would take ten children in the opposite direction.

Norman Borlaug took his place in the middle of the line of children. It was a law with them when the snows came.

The oldest, strongest boys would lead the file—heading into the thick, crisp snow, searching to avoid the deep drifts, breaking the trail—while the oldest girls brought up the rear and kept watch over the young in the middle of the line. The bigger boys and girls knew the law and the reasoning behind it: getting to and from school was a part of growing up, a way to self-reliance. Only in the worst blizzard conditions would one of the fathers hitch up a team and sled to come for them.

There were about twenty children in the straggling line, trudging to the group of farms near the little village of Saude. Norman Borlaug was the smallest and youngest and this was the first time he had made the long journey through the snow.

The road would take the children straight ahead for two miles, the second mile slightly downhill and more exposed. Then they would turn right at a crossroad and walk another mile to the gateway of Norman's grandfather's farm. But even before they reached the crossroad, the girls at the back were calling to the leading boys to slow down. The smaller children could not keep pace with them; they were struggling. Snow clung to their clothing, faces, and hands, and the frozen surface came up to Norman Borlaug's thighs with each step. Snow melted inside his boots and soaked his stockings; his legs went numb with the cold. He began falling often. Soon the challenge was too much. Each lunge forward was more exhausting than the last. He gave up. Filled with utter weariness, an embracing tiredness that stopped him in his tracks, he sat down in the snow, cuddled his frozen knees to his chin, and tried to cry himself to sleep.

A hand pulled his scarf away, caught his fair hair, and yanked his head backward; the face above him was tight-lipped with anger and fright. It was a cousin from his father's side. "Get up!" she yelled at him. "Get up!" When he did not move she began to slap his face, striking him with open-handed blows—backhand, forehand—again and again. "Get up! Get up!" The other children gathered around them, calling to him with fear in their voices. Some of the young ones were crying.

The punishment was stopped by one of the big boys. Norman was lifted to his feet; the snow was brushed from his clothing and his legs. His cousin took his hand and helped him start walking again. She was ten, self-assured and full of authority. He was naughty, and stupid, to think he could go to sleep in the snow. He hoped she would not tell his grandfather. He felt guilty because what he had done was not what his grandfather called "common sense." The tears froze on his face.

He was still crying when he came into the warm farm kitchen. His grandmother had just taken the loaves from the glowing Norwegian-style oven and the house was full of the lovely aroma of fresh-baked bread. She always baked their bread in the afternoons so that its fragrance would be a welcome for returning members of the family. He was much older before he understood this, but for the rest of his life no homemade bread ever smelled as sweet to him as the loaves his grandmother baked that day.

This was the first time Norman Borlaug knew what it was to feel the pain and indignity of open punishment. At the moment it was secondary to the agony he felt in the warm kitchen when the blood began surging back into his frozen hands and feet. He cried and cried. His mother put his numbed hands under her arms, held him to her breast, and kissed his sore red cheeks, consoling him with her soft voice. His grandfather came with a hot, rough towel and rubbed his feet until the pain stopped.

He sat at the table to eat and saw the butter melt and the stiff, cold honey run thin on slices of hot bread. He sipped a mug of warm milk his mother put into his hand. He had stopped crying but still felt sorry for himself. His three-year-old sister, Palma, took no notice of him but played with a toy on the floor near where his five-month-old sister, Charlotte, was sleeping in the wooden rocking bed. He ate his bread and butter and honey, and drank his milk; and when his mother asked if he felt better he told her that his cousin had slapped his face. His grandfather picked him up and carried him to his seat by the open fireplace, warming his bare toes in the heat, speaking to him in his deep voice, pleasant and persuasive, with the musical lilt of his Norwegian origins.

"That girl was not bad on you, Norm boy"—always Norm boy or Norm, never Norman. "She did what she'd bin told and that's common sense. She did right—she's a good girl and you'd be dead pretty damn quick if she left you sleeping in that snow. She made you wake and walk, and that was good."

They put him to bed earlier than usual that night.

He lay in the attic room under the wooden shingle roof of the house that his grandfather had built long before he was born, remembering the old man's words. Common sense was always right. He wished he had a brother—older or younger, it wouldn't matter—to talk to about today and what happened in the snow. His face still smarted from the ice and wind, and from the slaps. He went to sleep wishing very hard that the winter would end and the spring would come.

The cold stayed all through that winter in northeast Iowa. But not all the days were dark and overcast. Sometimes the trees along the road to the schoolhouse, and along the small river that the children crossed by a wooden bridge, were dressed in white and coldly beautiful, their snowcaps and icicles catching a beam from a low sun and reflecting the light with diamond brilliance. On these days the Indian file walks became parades of laughter, and in the mornings Norman would run from the house, with its smell of frying bacon, to join the children gathering in the roadway.

Finally the belt of frozen air retreated, back over the northern counties, across the Great Lakes and the white expanse of Canada; spring was heralded with the crackle and jostle of the ice breaking up in the river. Nipping frosts relaxed their grip slowly, unwillingly. The still, cold silence of winter in the clumps of elm and oak and among the willows and poplars that lined the streams and the river gave way to the sounds of falling ice and dripping water, and to the song of the returning birds. These were sounds of a changing season that Norman Borlaug came to know and long for. The hard winter done, he'd walk freely home from school, across the wooden bridge, stopping to look down at the jagged ice slabs. Some days he wandered along the banks of the river to a sharp bend where the chunks of ice jammed together and piled one on the other before crashing onward with the speeding water. He knew the ice would soon melt, that the water would flow on into a great, warm river called the Mississippi and would empty out into the Gulf of Mexico. He knew this because his grandfather had told him. He could not imagine what it was like there, because no one in the family had been that far from home. It seemed as distant as another world.

The days of spring were busy on the 120-acre farm. These were the weeks when everything sprang to life and activity. The house changed from a closed haven to an open place where his mother cleaned and washed, her singing voice coming through open windows. The lines full of flapping linen; his sister, Palma, playing in the open air. The fruit trees—pruned so long ago—popping buds, and the willows with their first sheen of gentle green. His quiet father, always hard at work, would be with his grandfather in the old shed, sharpening tools, cleaning machinery for the coming work in the fields: the planting of the summer oat crop. Soon after Norman's sixth birthday, on March 24, a dancing, golden day when the sun held the first real warmth of spring, his grandmother went to the old log cabin, where her husband had lived as a boy and which now was used for storing utensils and for smoking hams in the fumes of oak. From the back wall she took down an old-fashioned fire-blackened cauldron and carried it into the sunshine in the yard. From the shadow at the back of the house he could see her white hair glinting in the sun as she carefully and firmly set the pot up on three flat rocks, leaving plenty of air space below it. She lifted her head and smiled at him, calling: "Fetch me the wax matches and the kindling, Norm boy! Light me a fire good. We have the soap to make this day for the family!"

She was bilingual. Like his grandfather, she had been born in America of Norwegian immigrants. Her family, the Swenumsons, had a farm three miles southeast of Saude, where they had settled soon after the Borlaugs came. Theirs was a closely related community. His own mother was from the Vaala family, the original settlers of the district. The older women on the farms had spent only a year or two at school, learning to read and write. They got their skills at home: the baking of delicious breads and cakes in wood-fired ovens, the preserving of fruit and berries, the domestic arts that wasted little—like making soap from fat saved carefully the winter through.

The bright flames licked around the blackened pot, and Norman watched as the old woman stirred in the soda and rendered down the fat. It became his favorite task: lighting fires to make the soap for the family for the year. He loved lighting fires for his grandfather too, to clear more land on which to sow grain, vegetable, and potato patches at the western side of the farmhouse, fires fed with piles of brushwood cut from the river banks the previous year and left to dry. Spread into clumps, the burning twigs and sticks gave off columns of bright blue smoke that bent in the breeze and left patches of gray ash and mounds of charcoal pellets. When the spring rains came spattering down into the soil, driving the debris of his fires into the earth, his grandfather would nod his head and his intense blue eyes would light with pleasure. A thing to remember about the earth, he said one rainy day, was that it was like a bank—you got back what you put in. Repay the soil for what had been taken away last year and the soil would always make another loan. The vegetables, the crops, the plants would grow quicker, and happier, and fatter. It was another of his lessons, and though Norman Borlaug did not understand the meaning at that time, he knew it was common sense. He learned many things from the old man, not only in their vegetable patch but also in the fields, where they spent more time in the summer when school was out.

On Sundays he went to the morning service at the Lutheran church in Saude. The whole family attended, although his grandfather was sometimes critical of the hypocrisy he found there. "Norm boy," he would say, "I get sick and tired of seeing what's going on in the churches, saying one thing and doing another." But in fact he was a good Lutheran, especially in later life, and always supported the church. The Borlaug family had helped to build the imposing white wooden edifice with its high bell tower. All the children had been christened there, and it was a rallying point, a weekly meeting place for the branches of the different families, people whose past, present, and future were linked to each other and the land. At the end of the service, when the men gathered outside, they talked seriously, not of the sermon and its solemn meaning but of the soil, the crops, the weather, and the harvest prospects.

After church the family gathered for lunch in the Borlaug

kitchen, his grandmother sitting at one end of the table, his grandfather at the other. Like all their meals, it started with his mother uttering a prayer of thanksgiving. Then there was the delight of the afternoon when his father and grandfather exchanged glances and went to the shed where they kept their fishing rods. Norman loved the sport; to sit by the men on the banks of the Turkey River and watch the lines go suddenly taut and see the silver catch twisting and wriggling through the air, glistening in the sun. And he loved too the way his mother cooked the fish in the oven for supper, baked in vinegar, with onions.

Summer brought him many good days, days free to wander, to see the work in the fields, to ask questions. But the happiest, most joyous time was when the harvesting machinery rolled down the roadway and turned into the farm gates. Rumbling, clanking, the old steam-fed traction engine led the way, its tall, cowled stack puffing blue-gray smoke; behind came the threshing machines and the wagons crammed with men and boys—uncles, brothers, fathers, sons—shouting back and forth, sitting on the piles of sacks that would be filled with grain by evening.

The community arrangement, in which the local farms owned and used the machinery on a cooperative basis, was that all able men and boys went from farm to farm with the machinery each year to help with the harvest. It made a festival of harvest time, an event when fellowship flowered: the voices of the men calling, joshing the women, and the women laughing and working to provide the best table in the district on that year's harvest round. The makeshift tables, planking on trestles with white bedsheets thrown over them, would be loaded with chickens, hams, sausages, salads, breads, pies, cakes, fruits, and homemade candies, all laid waiting and ready at the back of the house.

Norman would be out in the fields, absorbed in the work—the great iron flywheel on the steam engine turning the long, drooping leather belt that powered the threshing machines; the cut chaff blowing into heaps; the shining grain filling the sacks, which were then lifted by strong arms into the horse-drawn wagons. Harvest coming in; hope fulfilled. His father and grandfather, eyes shining. Activity all day, and many different voices: some from the Irish community out toward the second nearest rail town of New Hampton; some from the Protivin and Spillville settlements of Czechs and Bohemians. People of the land, settlers in the new world. That spirit was caught by Dvořák, during two summers he spent at Spillville, where he composed part of the New World Symphony.

There was singing and drinking when the work was done, the oats harvested, the small wheat crop for the family stored in the shed. One of the songs told of how the wheat and corn grew taller in Iowa. Norman could not understand why they did not sing about the oats instead. When he asked about it, his grandfather told him that wheat used to cover all the acreage in past years, but that something had gone from the soil and now oats did better. No one knew why.

Later, when they lifted the vegetable root crops—the carrots, turnips, parsnips, and potatoes—he thought about the soil and the wheat and the oats. He asked his grandfather whether they could grow wheat again if they burned enough brushwood on the fields. His grandfather smiled at the idea.

They tied the onions into strings and hung them on nails high on the walls of the old log cabin. Carrots and parsnips were cleaned and stored in a pile of dry sand. Potatoes were washed and stacked in bags on a raised wooden platform so that they would get air and not go moldy too soon. Apples were carried in from the orchard, then wrapped in pieces of paper and packed in a big barrel in the room across from Norman's attic bedroom. In the kitchen the table was filled with jars and bottles, and pans steamed on the stove with fruits for jellies and conserves.

It was autumn suddenly, and they were preparing for winter. School started again, a full winter classroom; and there were new children. He was a year older, no longer the youngest or the smallest, and when the first snows fell he knew how to behave with common sense when they formed the line for the Indian file walk home.

In April 1921, in the flourish of the spring, Norman Borlaug had his first encounter with death. His mother was expecting another child, but he was told nothing of the impending event. He had just turned seven, and in those days a boy of seven was not expected to understand the biology of human birth—though he might see reproduction in many forms all around him on the farm. He was oblivious to the unusual activity in the farmhouse: the comings and goings of serious-faced relatives, the doctor's frequent visits. His first indication of anything unusual came on the morning the child was born.

His father woke him in the attic bedroom and told him he would not go to school that day but would stay home to help with the younger children. Why? Another baby had joined the family, his father said.

Was it a baby brother?

No, it was a girl, another sister. She would be called Helen, and he would see her later in the day. His father looked sad; he did not know why.

He jumped out of bed, his disappointment in the sex of the new arrival fading in the disruption of routine, the atmosphere of portent. Late in the day they allowed him to see his mother. She was sad and quiet, with moist eyes. He saw the baby in the wooden cot, sickly and silent, its eyes closed. He never heard it cry.

For four days gloom hung over the house, and on the fifth

day he saw his grandmother weeping in the kitchen and his grandfather comforting her. Then his father came and took his hand and walked him out into the sunshine of the yard. He said the new baby had "gone back to heaven"; that he and his two sisters would go to stay a few days with Grandad Vaala because there were certain things to be done. His father gave no reason why the baby girl had gone to heaven, but later Norman was told there had been no hope from birth. When the baby had been buried in the grounds of the Lutheran church in Saude, he and his sisters came home. His mother kissed him, full of tears; his father smiled again, and his grandfather took him fishing.

After a day or two life returned to normal. Norman's sister Palma joined the rural school after Easter and he had new responsibility as her guardian on the long journey each day. The disturbance caused by the tragedy was past and seemed to be forgotten. Not until he was much older did he know of the deeper consequences of the event—the medical opinion that Clara Borlaug had little hope of bearing more children. The only sign of change was in his grandfather's attitude.

The old man understood his grandson's longing for a brother, and he knew that the longing would never be met. From the time of the baby's death he shared as many hours of the day as possible with the boy, bending his white head to talk to him with homespun philosophy, recounting tales of the past, of places and the lore of the land.

Three times a week, after school in the early evening, his grandfather took him into Saude to deliver produce. The village, settled in the 1850s with a few log cabins, now boasted a general store for supplies, a creamery that purchased the farmers' surplus milk products, a feedmill that ground wheat to flour for breadmaking, and a blacksmith's shop with a forge and big bellows that sent sparks jumping over the stone floor. On one side of the street the residents had built a footwalk of wooden planks so that the women would not have to drag their long skirts through mud and dust.

Occasionally on Saturdays they would ride into the nearest rail town of Cresco or visit friends in the Bohemian settlement at Spillville, fifteen miles away. The settlers there continued to preserve their culture and dressed in colorful national dress, the girls with pretty puffed sleeves and embroidered aprons, the men with handlebar mustaches and long, curving pipes with lids that flipped open to light the tobacco. It was a small, gay township, and it was there young Norman learned to dance the polka.

During their times together his grandfather revealed a cast of mind that was so free-ranging and questioning that Norman Borlaug later came to regard his grandfather as a born rebel. The old man argued that ideas, habits, behavior were not necessarily right merely because they were widely accepted. A man had to think for himself. Custom, tradition, bigoted religion—all were hotly challenged in his earthy manner.

"Don't make sense to take things at face value, Norm," he once said. "Got to keep your own mind for yourself—take a good look and weigh things up; don't get jumping to conclusions in a hurry. Most times you'll be sorry in a hurry if you do. Take a good look first. Same with people as with things. Look carefully; you'll soon pick the ones that's solid. Watch out for those that do all the talking like they was on personal terms with Him—never doing a tack of God's real work, like growing things, or making things; just sermonizing, gossiping, and scandalizing. You watch them—and keep clear."

That summer there was another morning that Norman Borlaug recalls vividly. He was with his grandfather and father, hoeing weeds between the potato rows after a shower of rain. White clouds floated high overhead, and birds moved across the sky like black smudges, waiting the chance to scratch in the fresh-turned earth for worms and insects. The air was still, heavy, and redolent with the rich, full scent of the soil. He clutched a handful of earth close to his face and sniffed the smell deep into his nostrils. His grandfather watched from the next row, motionless over his hoe, blue eyes beaming with some remembered pleasure, intent on the boy's face.

"Soil smell good to you, Norm boy? Sniff it hard. That's life you're smelling; that's the smell of land, and man is lost without land—remember that! Soil makes grass, every blade of it; and grass puts flesh on cattle, makes milk, butter, cheese, feeds flowers the honey comes from, feeds acres of oats and wheat that give us cereals and bread. You see those crafty people looking pious up at the sky, son, you tell them to look down here. That's where you find God—in the soil, in growing things."

The sight of his grandson smelling the earth brought the old man memories of his own boyhood on the same land. For more than sixty years Nels Borlaug had known the scent of the Iowa soil around Saude—after rain, and when it was baked dry in the hot summers of drought and filled the nose with dust. It had given hope to his immigrant parents, had nourished four generations of the Borlaug family. His attachment to the land was deep and fundamental; but he was conscious too of the roots of his people across the seas. He talked often of these things to his grandson, going back, as old men do, to the past—to when his people came across the ocean looking for a new life and fertile land, to when his parents were pioneers and there was danger in the West.

Nels Borlaug's parents had left their home near Sogne Fjord, in central Norway, in the early 1850s. Worried by signs of upheaval in Europe, and struggling to stay alive on a few acres of poor land, they were ready to face the unknown and to endure hardship rather than live without hope of improvement in their own country. They took their two young boys, Nels' older brothers, and sailed across the Atlantic with a group of other Norwegian families. At the same time Irish emigrants were streaming across the ocean to America in the tens of thousands, fleeing the fearful famines that had followed successive potato blights in Ireland. So many were sick and penniless that they had created enormous problems for the immigration authorities in the Atlantic ports. Consequently, the Norwegian group took a different route to avoid difficulty. They went up the St. Lawrence River to Montreal and from there by way of the Great Lakes to the shores of Lake Michigan, near what is now Milwaukee, Wisconsin.

The Borlaugs had teamed up with several other families on a trek across country and settled on land near Madison. The immigrant families formed close friendships, giving help and encouragement to one another in their new country. Among those with whom the Borlaugs had a particularly close relationship were the Vaalas and their relations, the Landwerks and Swenumsons, from southern Norway.

The first year brought the expected hardships; and in that year Nels Borlaug was born. Before he was a year old his father and mother packed their belongings and the children into a covered wagon and set forth looking for a homestead far to the west. They found the green pastures, near Vermillion, South Dakota—but not the prosperity or the peace of which they had dreamed. They spent a perilous year perched on land bordering the wide Missouri River, in perpetual danger from attacks by the Sioux, who were raiding isolated and unprotected settlements. Nels' father never moved without a rifle close at hand. He could not go far from the cabin he'd built for his wife and three sons, and it was more than a year before relief came by way of a letter from the Vaala family.

The Vaalas and the Landwerks had left Madison shortly after the Borlaugs and had struck due west toward Cedar River, Iowa. They reported in their letter that they had come upon promising land open for settlement—good, flat soil close to water and timber—and that there was much more available if the Borlaugs wanted to join them.

By the time the Borlaugs arrived, the two families had built cabins and named their little settlement Saude, after their original home near Telemark. Other families also began arriving, among them the Swenumsons, whose daughter Emma would become Nels Borlaug's bride.

Nels could not remember his father and friends building the log cabin in which he lived as a boy, but it still stood in the 1920s, rock-firm as the day it was erected. His eldest brother inherited the original holding, and Nels took up his own 120-acre farm adjacent to the old homestead. It was there that Norman Borlaug's father, Henry, was born, the second of Nels and Emma Borlaug's four children. Henry went to the small rural school, as had his father. But his father had stayed for only three years, leaving as soon as he could "read, write, and figure," which was standard for the hard-working farmers on the prairies. Schooling was a luxury; high schools were part of another world and out of reach of all but a few.

Nels Borlaug was left with a deep hunger for the education that circumstances had denied him and was determined that his son Henry should be given a better chance. So although they were barely a step ahead of poverty and could have used an extra pair of hands on the farm, Nels sent his son through all eight grades of the country school—and even managed to put him through two winter courses in a small business school in Decorah, thirty miles to the east. But by the end of his second term Henry had met Clara Vaala and had fallen in love. They married and settled down under Nels Borlaug's roof.

Clara Borlaug was a small woman, very shy, and an excellent cook. Like Norman's grandmother, she loved to bake bread. She made wonderful cakes and fruit pies, and Norman and his sisters would go to the woods at the back of the farm to gather wild blackberries, raspberries, and cherries for her. In the fall she would start the canning, boiling the mason jars in big pots on the old stove. She canned everything—chicken, corn, beets, peas, carrots—and stored the jars in sand in the cellar. The kitchen was a place of perpetual motion and tantalizing aromas.

Norman's father was six-feet one, solidly built, and an indefatigable worker. He spent incredibly long hours in the fields—in spring, summer, and fall it was sunup to sundown—so Norman grew up under the eye of his grandfather, rugged and grizzled, the patriarch of the family. The old man held to his conviction that learning made the mind strong and effective and took every opportunity to demonstrate it to his grandson. There were times during their trips to Saude when they would sit in the cart in the shade and he would point out certain people in the street.

"Now see that young feller there? See how he moves? Sure of himself, certain and positive what he does. You watch him, Norm boy. Got the best education, he did. Education and common sense, thinking things out for yourself—that's what matters."

The implication was always clear. Follow the example. Despite his lack of formal education, the old man was widely read and wrote in a slow, but meticulous hand. He kept his account books in the same careful way. One day he tossed the cashbook across the table to his grandson.

"Here, Norm. Your teacher says you've done well at arithmetic—see if I made any mistakes in my figuring in there." Norman was eight years old, and the device was to get his mind working, but the cashbook revealed an economy of subsistence he never forgot. The carefully penned, well-set-out figures showed that the family income for the year had been $128—$2.50 a week for a family of seven, with four adults at work the year round. The spending was limited to essentials such as:

Salt—2 cents
Pepper—2 cents
Vinegar—4 cents
Cotton, needles—5 cents
Four yards of red flannel—$1
Can of oil—20 cents
Whiskey—$3

Items like work boots were made to last for years, stitched and rubbed with fat, tacked and resoled until they would no longer hang together. The children's clothes were sewn patiently in the long evenings, stitched by the light of an oil lamp. Dresses, shirts, trousers were all hand-me-downs; bedding was carefully patched and nurtured against hard washing or hard wear. Curtains were few and treasured, a luxury prized by the women. It was a frugal life; but if they were poor, the children were never hungry. There was not a day in his life when Norman Borlaug did not have something to eat. The home-baked bread was always there, with the butter and milk from the cows. There was honey from the hives, fruits and vegetables from the garden, ham, poultry, and other meat.

But—more than a week's income for whiskey? He asked his grandfather why so much went on this drink.

"Could be vital, Norm. Might mean life or death to have it on the spot when you need it, boy. And—I can tell you—it's often saved me a long, expensive ride to the quack in Cresco!"

He remembered the words the following winter, when he saw a new side to his grandfather. A few of the old man's cronies came one day, among them cousins from the other families, riding in a sled drawn by two big, strong horses, with everyone laughing and talking. Their guns across their knees, they went off on their "hunting trip," looking for rabbits, squirrels, and other game.

Norman was in bed asleep when the returning hunters came roaring home, stomping into the house with snow on their fur caps and on their boots. He crept out of bed, sat halfway down the stairs, and saw his grandfather take the bottle from the cupboard and make a hot toddy from the steaming kettle at the fire. They were big, hearty men, and they said a lot of things and laughed at a lot of jokes he did not understand; and when they became merry, they sang Norwegian songs learned from their fathers and mothers.

There were other times when the Vaalas visited them, or the Swenumsons, and the bottle was passed around and there were more jokes and songs. But he never saw his grandfather drunk—only merry. Then he seemed younger, and different. Even so, Norman Borlaug had not seen any life-or-death reason for spending a week's income on the bottle.

In the summer of 1922, when Norman was eight, there was more upheaval and change in his life. After much discussion it was decided that his father should buy his own farm—a fifty-six-acre holding that touched on one corner of the Borlaug land. Nels Borlaug argued that the land was cheap and would soon get much dearer; that it would increase the family holding; and that if the size and the soil proved too unproductive to support Henry Borlaug and his family, they could consider merging the two farms and living and working together again. The old man's arguments won the day. He raided his life savings of $800 and placed his own farm deeds as collateral with the bank. He would have to work harder in the fields and find help, but it was time Henry and Clara had their own home. After all, a man as healthy as he was could live to see a hundred years. They just couldn't wait that long to inherit.

The first few years on his own farm produced more challenge than profit for Henry Borlaug and his family. The land was wooded—mostly oak—and had to be cleared. The soil was variable and in places quite different from that on his father's farm. Crops did well in one spot, but not in another. Chemical fertilizers were virtually unknown. Barnyard manure, though effective, was scarce.

The new home also brought challenge to Norman Borlaug, with longer hours of work with his father and a more demanding role at school now that he was in the higher grades. His grandfather paid them regular visits and watched his grandson closely. The boy was growing lithe and strong, with his arms strengthening and shoulders filling out from digging and hoeing, lifting bags of oats, and wielding an ax to split wood for the fire or fenceposts. He was gratified that the boy reveled in the outdoors and the work, but it also worried him. Education was too important. When harvest time came around, he watched his grandson riding on the wagons behind the clanking machinery with his father, cousins, uncles, and friends, singing on the way to the next farm—and there was every appearance of a lad bound for life on the land.

When Norman asked his father if he could stay home from school to help on the farm, Henry seemed touched by the boy's eagerness but would not give his consent. His own views coincided with Nels'. Henry used few words to express himself, but he liked books, and he read everything he could get his hands on. Norman Borlaug recalls the time there was a tornado and the family had taken refuge in the cellar. His father was pressed against the door reading *The Saturday Evening Post* by candlelight. He was leaning on the door to keep it from blowing in.

In his own quiet way Henry Borlaug told his son that the

loss he would suffer by missing school would be far greater than any gain from work in the fields. Norman was disappointed. But later his grandfather spoke to him when he was alone: "When you're older you'll understand, Norm. You're wiser to fill your head now if you want to fill your belly later on."

The old man intensified his crusade. He bought books, picture magazines, and newspapers. "Here, Norm boy," he would say, tossing a newspaper across his knees. "Read me that article; my eyes are not so good." All the time he worked at raising the boy's sights, beyond Saude and Cresco, lifting his mind to the great continent and all its problems and promises.

One evening he walked into their kitchen carrying a strange-looking box with a long trail of wires. "Just wanted to show you this," he said, tapping the box. "Got it from a feller in Cresco today—crystal radio he calls it. Come outside and listen." They went outside into the cold air and the old man threw a long wire up in a tree to improve the reception of the primitive radio, took earphones from his pocket, and listened while he fiddled with the box. He put the earphones on the boy's head and saw his bemused, entranced face. Norman stood unmoving in the cold air, his mind caught in the magic of the moment—music and strange voices through the night, across the continent, sounds from a world beyond his experience.

But his curiosity was sparked by the things around him as well. His father and grandfather were assailed with a running stream of questions that came from his thirst for information. Why do those trees grow straight and tall and these bent and thick? What bird is that calling? Why do ants climb trees? What does the fungus growing on trees live on? Animals, plants, insects, birds gripped his imagination. Why do mushrooms and toadstools grow with such a rush?

One day in late summer, when the oats and wheat were ripening and they inspected the crop, the boy felt keenly his father's disappointment: stands of good grain in one place, poor heads in another. Crops had succeeded and crops had failed, all on the same piece of land—all without explanation.

He was fourteen now, and here was the same mystery of the soil he'd seen on his grandfather's farm. Burning brushwood helped in one place but not in another. Fertilizer was only partly successful. And there was not a real clue unless it lay in his grandfather's remark that many years ago wheat had been grown on the same land year after year until the yields were so poor it had been turned over to oats and corn. Crop rotations with legumes had improved the fertility of the soil and increased yield considerably—but not impressively. Something more was needed.

The boy voiced the question in all their minds. Something is missing. What is it? He did not know he was probing into a realm of science not yet fully developed, concerning the interrelationship between organic and chemical life in the soil. It would be some time before scientists would demonstrate that successive generations of the same crop sap the land of vital nitrogen, potassium, potash, calcium, phosphorus, and rare minerals necessary for productive soil. He could not know that under his feet, under his father's work boots, was a phenomenon that plagued the ancient agriculture of impoverished nations. He could only ask questions. And there was no one who could tell him the answer.

Later he sat with his beloved grandfather by the river, fishing quietly, not talking. Suddenly the old man broke their silence.

"Just don't stop asking why things are the way they are, Norm. One day someone'll tell you the answer."

Two

During Norman Borlaug's final year at the rural school the family decided he should go to high school. His eighth-grade teacher, Sina Borlaug, was instrumental in the decision. Because she was a relative, a second cousin on the Borlaug side, her argument carried great weight in the family conference. "No question! He's no great shakes as a scholar; his arithmetic is awful—but he sticks. He's got grit! High school will make him."

So early one September morning in 1928 his parents walked him to the farm gate to see him off, and with his back straight as a rod he strode briskly the mile of dirt road to the Cresco-Protivin junction to wait for the farmer with the Model T Ford who would pick up Norman and other students along the line. The parents took turns driving the children to school—except in the dead of winter, when the roads became impassable with deep snow. Then the students would stay for a few weeks with families in town.

Norman adjusted quickly to his new surroundings and was soon caught up in a larger world of sports and scholastics. Classes were oriented toward agriculture, and the major course of study, the Smith-Hughes agriculture program, was considered to be a terminal preparation for boys going back to live on the farms. The man who taught the program, Harry Shroeder, had a keen scientific mind, and it was he who first sensed Borlaug's innate curiosity about the processes of plant growth and the nature of soils. As the youngster greedily absorbed information, Shroeder sensed a quality in the boy that went far deeper than rote learning, and he began tutoring Borlaug on everything he knew about agriculture.

Another teacher who was to have profound impact on Borlaug's life was Cresco's principal, David Bartelma, who served as the school's athletic coach as well. A short, husky man in his late twenties, with an erect bearing and strong chin, Bartelma gave everyone the impression that he was much bigger than his 145-pound frame indicated. In 1924 he had gone to the Olympic Games as an alternate on the American wrestling team, and to Borlaug and the other boys he was an imposing figure. Bartelma was the driving force behind Cresco's educational program. He believed that the most dynamic force in man's character was the competitive spirit: that if a boy gave his best on the playing field, he would give it in the classroom and in his life. He was always trying to help boys with character and intelligence, boys who would repay special counseling with their pursuit of higher education—and competition in sport.

One afternoon, as Borlaug was playing football and fighting furiously against linemen who were older and stronger than he was, Bartelma called him over to the sidelines.

"I want you to join the wrestling squad and wrestle for the school," Bartelma told the astonished boy. "You're quick and strong for your size and you should do well. But you'll have to give it all you've got or you'll be out."

The invitation was brusque, but Bartelma had already turned Cresco into a fervid center for wrestling, and the boys he trained would star later on the varsities of the Big Ten universities and even in the Olympics. Borlaug felt honored to be given the chance to join the team, especially since he was small for football. It was his toughness and courage—attributes that were vital in the lighter weight classes of the wrestling team—that had brought him to Bartelma's attention.

As Borlaug began a regime of special training, he first tried to overpower the coach in practice sessions. When that didn't work, he tried to out think him, using every possible combination of the new holds he was learning. Whenever this happened Bartelma would stop the practice.

"You're thinking way ahead of yourself and you're not coordinating your thoughts to your actions," Bartelma would caution. "If you're going to be a good wrestler, you have to discipline yourself to think and act at the same time."

Bartelma's coaching achieved the desired results. Borlaug became one of the school's best athletes—a champion wrestler and a star football and baseball player. But Bartelma's philosophy produced even more. He wanted everyone's best: each according to his ability, according to the situation. "Give the best that God gave you," he would say. "If you won't do that, don't bother to compete." This was his code—for sport, for study, for work, for life—and it produced a lasting bond between Borlaug and Bartelma.

During the next year Borlaug found it easy to make new friends. Some of them visited the Borlaug farms during vacation, enjoying haymaking, harvesting, botanical hunts through the woods for little-known plants and insects, or riding to Spillville on Saturday nights to dance the polka with the girls of the Bohemian community. And as the summer of 1929 drew to a close, it looked like a good year for the Borlaugs. Norman's father was doing well with his farm, paying off the mortgage and saving money against the day his son might go to a university. Both farms were doing well, but his grandfather was working less and less, using more hired help. At the end of the bountiful harvest they celebrated his seventieth birthday, and the farmhouse where Borlaug grew up was filled with relatives from both sides of the family, all prospering to a greater or lesser degree since the wagons had rolled west across the prairie and the Vaalas and Landwerks had founded Saude in the previous century. They made a fuss over the old man; he was slower and showing his age. He grumbled at the nonsense, the breaking of routine, but he enjoyed himself. As the gathering grew merry, some relatives joked with Nels Borlaug and said they promised to come and toast him when "his hundred came up." It seemed a good end to a good summer, though not one of the celebrators could know that an era was over.

Late in October the New York stock market crashed, plunging the nation into the greatest financial disaster in American history. The collapse wiped out billions of dollars' worth of investments, and the reverberations shook the Western world into what became known as the Great Depression. It did not strike the Midwest with the same suddenness as it did urban centers. There were none of the scenes that appeared overnight in the city streets: long, patient queues of ragged men waiting in the cold at the steaming soup kitchens; hundreds of people clamoring for a single job. But while the Depression was not as visible in the country, its effects were being felt as more and more farmers were driven into a frugal existence in which they lived with the fear of dispossession, foreclosure, and bankruptcy. Seeds became as precious as in the old pioneer days; crops were guarded against theft by hungry marauders; the sky was watched intensely for signs of drought, which would wipe away all remaining collateral and all hope.

It was becoming harder for the Borlaugs too. Norman could see that the times were rapidly aging his grandfather, while his father kept his counsel and said little. Fortunately, he had purchased the lumber for a new barn and it had arrived just before his bank had closed its doors—like many others—with the remainder of his cash reserves. It immediately became a sacrifice to keep his son in school, but it was a sacrifice he was determined to make.

When Borlaug's class at Cresco High School graduated in 1932 his schoolmates went their separate ways, each with the face of the hard times looking over his shoulder.

Norman Borlaug had a fine scholastic and sports record; he had been named one of the school's outstanding students. But that was all. There were no awards, grants, or scholarships. Worried about his future, he talked with his adviser, Harry Shroeder, who held out one hope—a prospect of an opening at Iowa State Teachers' College in Cedar Falls. There was one scholarship to train as a science teacher, but it would not be available until the following year. It was not a university, but it was his only chance for a college education and he decided to apply and wait out the year.

He was eighteen, competent, and full of energy, so the wait would not be a hardship. He could spend the time helping his father and grandfather and, when they could spare him, working as a hired hand on local farms for a few cents an hour—pittances that he would hoard away for what might come.

When the summer was over and the fall harvest in, work was harder to get. The long winter days of inactivity after his full school life made him restless. He cut fenceposts for thirty-five cents a day in weather twenty to thirty degrees below zero. He started hunting and trapping, tramping through the snowy woods, setting snares for pelts he could sell in Cresco or Protivin: weasel, squirrel, hare, muskrat, an occasional skunk, and even a rare mink, one of which brought him a fortune of six dollars. He went out every day to check his traps so the animals would not be left in pain.

It was hard work for the meager income it brought, but the meat the trapline produced also added protein to the family diet.

One particularly cold winter day he went further afield than usual, following tracks through the snow, along the tree-lined banks of the frozen streams and the river. It was already ten degrees below zero and turning colder when the light began to fade. He stumbled home in the dark, his clothing stiff with ice, his bones chilled to the very marrow. When he finally reached the warmth of the farmhouse, there was no feeling in his limbs, and with the returning blood he suffered the same agony he had felt that day so long ago when he had journeyed home from school Indian file. This time, however, the exposure had been more extreme. By the next morning he was suffering from severe pneumonia. Time became a blurred image of movement and voices. Days slipped from his memory. There was a vague recollection of his grandfather, with the old toddy bottle, rubbing his sore chest with whiskey, speaking soothingly in his deep voice while the pungent smell permeated the room. "Might be vital to life," he remembered the old man had said—long ago.

He recovered slowly. It was too risky for him to hunt any more during the winter, but when spring came he was well enough to begin hiring out to local farms again. He earned from fifty cents to a dollar a day, and by early summer he had hoarded a total of nearly fifty dollars. His father offered to double the amount if he enrolled at the university. He was tempted, but he argued that the money was needed for his sisters, Palma and Charlotte, who were also eager to study. He would work and pay his way; they could not.

At long last a letter came from Cedar Falls offering Norman a partial grant at Iowa State Teachers' College, with the assurance that work would be found to meet his other expenses. He wrote back to say he was grateful, and that he would present himself when school opened in September. So his future was settled, but he was not really satisfied. He kept thinking of his grandfather's admonition to get a university education. "Get there, Norm, any way you can," he had said. "You ain't got any protection till you've got a university education. Sure, it'll be hard. You'll want to back off, but that's the reason for sticking." Now he wondered if he shouldn't have accepted his father's offer.

A break came unexpectedly—a little more than a week before he was due at Cedar Falls—when George Champlin came driving down the road in his Chevrolet. Champlin had also gone to Cresco High School but had graduated the year before Borlaug had entered as a freshman.

"You probably don't know me," said Champlin. "But I know you. I've just been talking with Dave Bartelma over at Cresco and I wondered what you were doing out here."

Of course Borlaug knew about him, though they had never really spoken. Champlin had been captain of the Cresco football team and was now a halfback at the University of Minnesota, but Borlaug could not imagine why he had sought him out.

As if to answer the unspoken question, Champlin came to the point. "I'm leaving today for early football practice and I'm also an unofficial scout. Bartelma told me about you. I'd like you to come back with me—today."

Whether Champlin had gone to Bartelma on his own or whether Bartelma had initiated the move, Borlaug never knew. In any case, there could have been no more persuasive scout than Champlin. He saw Borlaug's hesitation and listened as he explained that he had already accepted an offer from Cedar Falls. Champlin then gave him a further inducement.

"Your wrestling teammate at Cresco—Erwin Upton—he's already packing his bag, waiting for us at his family's farm on the other side of town." It would be companionship.

Borlaug still would not be hurried; it was always difficult for him to change his mind. There were problems of money and accommodations, he said. Champlin threw each argument aside.

Money? Find work like thousands of kids did these days. Accommodations? No problem. Champlin had a place in a rooming house in southeast Minneapolis, a single room with two double beds. Four of them would share the total rent of twenty dollars a month.

And as for Iowa State Teachers' College: "Make up your mind, then call them on the telephone; explain." He made it sound so easy.

"Tell you what," said George. "Ask your folks, pack your bag, and drive up there. You can see for yourself. You've got ten days before you go to Cedar Falls. If you don't like it, hitchhike back. It won't cost you a thing but time. Then you'll know for sure."

It sounded perfect. Borlaug grinned his acceptance and felt a welling of elation.

His parents took the change of plans without any show of excitement, but his grandfather was obviously thrilled. He walked with his grandson arm in arm out through the gate, where Champlin waited. Before the boys drove away, he took out his old leather wallet and thrust its entire contents of eleven one-dollar bills into Norman Borlaug's hand. "You'll make more use of 'em than I will, Norm boy."

He now had a little more than sixty dollars with which to start his academic career. It was a disturbingly small sum as he ran his fingers through the bills, counting in his mind while Champlin watched him with a smile. Tuition fees for the first quarter alone would be twenty-five dollars, and the margin left was thin. But Champlin was confident and eager to set his companion's mind at ease.

"You'll have no trouble. We'll get you a job for your meals and you'll be O.K Three-quarters of the kids at

Minnesota work their way through these days. There's always a door open. You'll see."

They picked up Borlaug's high school friend Erwin Upton, and the Chevrolet sped the fifteen miles to the Iowa border, then on to the twin cities of Minneapolis and St. Paul, 150 miles away. George Champlin entertained his young recruits with tales of campus society, of lectures and teaching and debates, and, of course, of the influence and power of sports.

Borlaug had no preparation for the impact of metropolitan Minneapolis. For a young man whose experience had been limited to rush hour in Cresco, the city was overpowering—in size, spread, ugliness, noise, and complexity. Street after dingy street, endless junctions and thoroughfares with streetcars and motor vehicles, rows of shops and then the dark gloom of shuttered restaurants and buildings—factories and stores—symbols of the cold hand of depression. There were figures huddled in doorways, under awnings. Uptown there was evidence of some prosperity—lighted windows, well-dressed people walking briskly.

But there was barely time to take it in. Champlin was hustling. The car turned into Fifteenth Street, four blocks from the campus, and stopped at the rooming house. They walked up the stairs to the second story and looked around the single room with its two double beds, a table, and four chairs—the place that would be home for the coming years. They dumped their bags and were hustled back downstairs to the automobile.

"We'll go straight to the coffee shop," said Champlin. "Can't hang around. The jobs are limited. In a few days hundreds of kids will be doing what you're doing now—looking for work."

The coffee shop was privately owned and spacious, with three counters and several booths and tables. It could seat about 150 people. A dozen students were needed to work there as waiters. The proprietor was busy and had little time to talk. Come in at seven in the morning, he said, wait on tables for the breakfast session. Work an hour, get a meal—three times a day. That was the pay.

Champlin took them back to the rooming house and bought them hamburgers on the way. He was pleased. "Well, fellas, you've got yourselves a place to eat and a place to sleep. You'll report for early football practice, pay your fees, and enroll in a few days—then you'll be in and everything will be O.K."

The next morning Norman Borlaug and Erwin Upton were up early and on their way to the coffee shop, just off the edge of the campus. Borlaug was nervous, fumbling even the simple orders as he ran in and out of the hot kitchen. The student manager of the restaurant was a huge man, a hammer thrower named Spens Holly, and he liked to run an orderly shop. Toward the end of mealtime George Champlin came in to see how things were going, and Borlaug overheard Spens saying slowly but with terrifying determination: "Either I will make a waiter out of him or I will kill him." The words sobered him in a hurry. Almost overnight he became adept at the job.

Working three times a day for his meals imposed a routine that continued for years. At six each morning he and Upton would climb out of bed and be ready to wait on tables at seven; they would return at noon for another hour and again at seven in the evening. Other student waiters came an hour later. It became a focal point around which their days revolved, and though the food they received was simple fare—and never oversatisfying—it was eaten with gusto. They sat at a long table with the other student waiters; the sessions were informative and often gay. The gathering sometimes included second- and third-year students who were experienced in the rigorous work-study routine and gave them helpful advice.

One morning shortly after he started Borlaug sat down to his usual breakfast—a dish of prunes, a cup of coffee, and a slice of toast—and noticed a pretty girl opposite him. She was bright and vivacious, with thick black hair, a generous mouth, and rippling laughter. Their eyes met and she smiled; he was at once attracted to her. Her gaze was calm and level, and she had the clean, fresh look of a country girl; most of all he liked her direct, unpretentious manner. She asked him what he planned to study and smiled when he said he did not know, but her smile did not make Borlaug feel foolish.

Margaret Gibson was twenty, a sophomore, and she seemed to him to be at ease with the world, sure of herself and blessed with good humor. She came from Scottish stock; her mother's parents had escaped hunger in Scotland toward the end of the last century, leaving Inverness for the biting cold of Canada before moving south across the border to the United States. Her father was a railroad section chief in Oklahoma, and she had grown up in the little wheat country town of Medford. They had no wealth and few possessions, but she, her sister, and her four brothers had all had a very happy childhood.

It had been at the urging of two of her brothers, George and Francis, that she came to Minnesota. The boys had left home earlier, worked their way north with the wheat harvests, and eventually enrolled at the university, where both of them played varsity football, with George becoming captain and all-American. They wrote enthusiastically of their life there, encouraging her to study. By that time her father was entitled to free rail travel for his family and was in a position to pay her tuition fees. So in 1931 she had entered the College of General Education and was studying child welfare. She earned room and board by waiting on tables, babysitting, and picking up other odd jobs in the evenings.

She was nearly two years older than Borlaug and seemed

to know all the ropes, but she also understood the shy farm boy from Iowa. They talked about their country backgrounds, and Borlaug found himself sitting next to her more and more often. They were comfortable with each other from the beginning.

Before classes started Borlaug and Erwin Upton decided to explore the city. Each day they would ride out in a different direction to the city limits on the streetcar for five cents and walk back to the rooming house in southeast Minneapolis.

Borlaug found the twin cities of Minneapolis and St. Paul confusing, a complex and jumbled area of nearly three-quarters of a million people that was still in the throes of forging a metropolitan character. There was the mixing and merging of cultures, and everywhere contradictions: the desperation of migrant groups; the drive and energy of men who hurried to their businesses; the apathy of the unemployed, who stood with hands outstretched. There were streets and residential areas that spoke of comfort and security and slums huddled around the dead hulks of factories that echoed poverty. There were enclaves of German, Polish, Dutch, and Czech, where men stood on street corners and lifted their voices in argument and debate. There were burly men, mostly Finns, from the closed mining camps of the iron ranges to the north, and angry Yugoslavs protesting the reduced working hours in the meat-packing plants of St. Paul. There were women in black shawls who spoke with the lilt and soft brogue of Galway and Connemarra, and so many fair-haired, serious-looking Scandinavians that the two cities together were sometimes referred to as Swede Town.

Here again was the pattern of the new world settlement Borlaug had known—Spillville and Protivin, New Hampton and Saude—but on a vaster scale, and with a depth of human misery he had never seen before: abject men with drawn faces and deep-sunken eyes standing in doorways clutching newspapers—not to read, but to wrap around their bodies against the cold of the night. There was hunger for food, hunger for work; and it troubled him.

There was another hunger in the metropolis—a widespread craving for culture and art and education. While the poor lined up for relief at soup kitchens, others lined up for tickets—to concerts by the Minneapolis Symphony Orchestra and choral groups of varying national backgrounds; to performances by the repertory theaters; and to the galleries and museums. The role of the university was strongly evident in the city's cultural life. It drew its students from every area of the state as well as from across the borders, and competition for admission was razor-keen. The senior educators were involved in more than running a university; they were providing a center for scholarship, spiritual refinement, and the arts of life. But they did not neglect the more practical aspects of education—the development of the skills and abilities that students needed to become active, productive members of society.

From his earliest days Borlaug heard the claim that this fusing made the university "truly universal"—that this combination of idealism and practicality gave the academic life of Minneapolis a decidedly humane character in which people mattered more than ideas or tradition.

Yet outside the university precincts he saw much that conflicted with this point of view. The university was an island in a sea of economic and social trouble. There were too many poor, too many hungry, too many aimless cabbages of men, too many people caught in the cold grip of a depression that the university could do nothing to loosen.

During his second week in the city he took the streetcar to visit the university's agriculture campus some five miles away, across the St. Paul boundary and on the edge of open country. He strolled across the complex, noting the locations of the various buildings and enjoying the pleasant country beyond. It was a gray, windy day, with a touch of autumn cold in the air, but he walked through the experimental plots to the woodland and the lakes beyond and wandered around for an hour. He liked the "farm campus," the peace and the air of the place. He decided he would like to study there—botany, zoology, forestry, plant pathology, soils, general agronomy, he didn't know which; but he felt this was where he belonged.

There was still more than two hours before his evening stint at the coffee shop, so he walked back to the main campus. He followed the streetcar tracks, across the St. Paul-Minneapolis city line, through a district he had noticed riding out earlier in the afternoon. It was a dismal spread of factories and assembly plants of World War I vintage. Now they were in decay and disuse, rows of them silent and shuttered. But the surrounding streets were lined with the little houses of the workers. He had seen them as the streetcar clattered through, but it had been quiet then.

Now he heard shouts, cries of anger. He turned a corner and saw a milling crowd jammed against the wire gates of a dairy products distributor—men and women, poorly dressed, screaming abuse at a line of big men standing menacingly on the other side of the wire with truncheons in their hands. A convoy of milk trucks, drivers in the cabs, waited behind them, and one of the men was shouting back at the mob, his voice lost in the uproar.

Borlaug crossed to the opposite side of the road and stood with his back against the wall of a deserted factory, wondering at the cause of the trouble. As he watched, the big men suddenly flung the wire gates open and charged the crowd, yelling and flailing about with their truncheons. There were screams of pain and rage; men and women fell bloodied to the ground. The crowd drew back and Borlaug was pinned against the wall, unable to move. He was not tall enough to see over the heads of the people; but he heard the engines roar as the trucks rolled into the street. Still there was shouting, blows, and threats. It was the most violent scene he had ever witnessed. The crowd ran after the retreating trucks, and he was free. He ran the opposite way, trembling from the fear caused by the naked display of hate and violence. There were people lying injured on the roadway, outside the wire gate, and no one tended them. The men still guarded the entrance with their truncheons held ready. A group of men and women still stood in front of them and yelled obscenities at the line of men: "Bastards! Bullies! Strikebreakers! Sons of bitches!"

The line of men stood waiting as the shouting went on.

Ambulances came, but there was no sign of the police. Borlaug got up the nerve to approach one of the older women. What had caused the riot? She looked at him, her teeth clenched and her eyes full of anger: "They cut the bloody wages in half—that's what! An' when the men went on strike, they brought in the scabs; and they brought in those gangsters to beat up our men when they protested. The bastards!"

The experience burned in his mind as he walked back to Fifteenth Street. All through the evening he recalled faces of brutal men, wildly swinging truncheons, and people falling. Above all, he remembered the fear he had felt.

Before he went to sleep that night he talked with George Champlin about his experience. But Champlin was not unduly troubled. It was the pattern of the times, he said. It was going on all across the country: riots, unrest, upheaval—they were symptomatic of the Depression. People got so little these days that taking away a bit more seemed a major loss to them.

Borlaug was sure it was more than that.

"None of us ever had much back in Iowa, George. You

know that as well as I do. But we never saw anything like this—ever!"

True, there were many more people, and greater pressures, in Minneapolis than in Cresco. But that wasn't all of it. What these people suffered was more than a loss of their income. They had become less than human beings. That was the basic factor behind the lean faces, the violence, the naked rage. He wanted desperately to talk about it, but his roommates had fallen asleep.

Staring into the dark that night, he knew with utter certainty that it had not been the brutality of the strikebreakers and their flailing truncheons or the madness of the crowd that had made him tremble and shiver. It had been the degradation of human beings, the display of anger caused by hunger. The scenes of that day were to stay with him always and influence his life's work.

When Norman Borlaug presented himself for enrollment, his academic career nearly ended before it began. He had gone to the university with Erwin Upton, and together they presented their application forms and high school qualifications to the evaluation panel. The knitted brows at once spoke of coming trouble. Normally students were admitted on the basis of satisfactory high school records, and no entrance examination was required. But they were told there was a difference between the high school systems of Minnesota and Iowa. Their freshman year at Cresco did not qualify as a first year of high school in Minnesota. They were a year short, on both science and mathematics. Consequently, they would have to take a written examination to assess their academic potential. They accepted that readily. At least the door was not closed to them.

Upton sailed through the examination and gained admission to the College of Liberal Arts. Borlaug failed. But he was granted an interview with the chairman of the evaluation panel, Dr. Ed Williamson. There was something about Borlaug that left Dr. Williamson loath to reject him.

The university had recently established a new institution for students who did not quite meet the entrance standards. It was called the General College. But it had quickly come to be regarded as a home for the misfits by the students and even some of the staff, who called it the College for Dumb-bells. Dr. Williamson, however, painted a different picture of the new institution.

"I am afraid we can't offer you any of the normal colleges, Borlaug; not with your qualifications. But this college offers a wide scope of subjects if you would like a place there: history, English, mathematics—on which you are poor—and science in various forms, as well as economics, psychology, basic health, and natural resources."

A sweep of subjects—but specialization in none. After two years a successful student would be awarded a degree of associate of arts. Borlaug was crestfallen, but he could not face total defeat and so accepted the offer. He went back alone to the rooming house.

That evening, as they sat eating their meal in the coffee shop, Upton and the other freshmen were jubilant. Borlaug was despondent. Margaret Gibson was sympathetic, and later for the first time he walked her home. Spilling out his disappointment, he talked with her outside the girls' boarding house, two blocks down from his room on Fifteenth Street, and she saw how intense he was, how keenly he wanted to do well.

"I'm nothing special," he told her, "but I know damn well I'm better than that—that place for the misfits they expect to drop out."

Margaret tried to heal his wounds. Might not his reaction be clouding his judgment? "Look," she suggested, "this could be something quite different, a chance to broaden your general education." At least he had been admitted to a university college—and whatever he decided to do eventually, he could only go up. It was a personal challenge.

He seemed so sincere; and she was concerned for him. "Tomorrow it won't seem so bad perhaps," she said reassuringly. "You might look back one day and say you had the wrong reaction to this." She was not to know it for years, but he was more sensitive than she knew and carried the pain of the lowly assessment for a long time.

Watching him walk away down the street, she saw dejection in his carriage and movement; and she wanted to hold and comfort him. Suddenly she realized there was more to her feelings than her natural reaction for people in trouble—and there was a sense of excitement in the realization. It was more than his blue eyes, more than the incredible honesty of purpose and drive that she saw in him. He had confided in her, and in a way that was flattering, more flattering than his obvious pleasure in her company.

She went to sleep that night looking forward to seeing him next morning at the coffee shop.

Although Borlaug's rancor was deeper than Margaret realized, he turned it into determination. From his first days in the new college he worked with an intensity that attracted attention. He was assigned to an adviser, Dr. Fred Hovde, a Rhodes scholar who was to become president of Purdue University. Fortunately, Hovde was a perceptive and understanding man. He saw Borlaug's ambition and sensed the rawboned strength of a country boy anxious to make his way. Dr. Hovde also watched Borlaug perform on the wrestling mat and baseball field and saw the same gritty determination to succeed.

At Champlin's urging Borlaug had gone to early football practice, but any hope he held of playing varsity football soon dissolved into realism. He was too small to compete against the 200-pound giants on the field and told Champlin: "They'd kill me in one play! I'll have to stick to wrestling and baseball."

The competition on the mat was fairer. The excellent training he and Upton had received from Bartelma at Cresco soon won them recognition. As the weeks went by they began forming the nucleus of a new force in university wrestling circles.

When Borlaug won his first intercollegiate match—wrestling in the 150-pound class against a freshman from Wisconsin—he decided to celebrate and asked Margaret to a movie. They agreed to forgo waiting on tables that night and to buy their own supper—a Dutch treat, except that Borlaug paid for the movie. They held hands that night in the theater.

The following week, during a chemistry lecture, he met a freshman named Scott Pauley, who was in the College of Agriculture, working on the campus Borlaug had visited the day of the milk riot. A likable young man with great flair and a ready wit, Pauley was soon a firm friend—of Margaret as well as Borlaug—and the three of them often walked into the country together on Sundays, looking at the trees and sitting and talking by the lakeside. Pauley spoke with enthusiasm of his studies on the agriculture campus, and his friendship reinforced Borlaug's determination to win a place for himself there.

He worked even harder at the General College, and by mid-December he had mustered the courage to knock on Dr. Hovde's door to demand that he be given a place elsewhere. Hovde was assistant dean as well as Borlaug's adviser, and needed some convincing. Why did he want to move on so soon? He had not yet been there a full quarter.

"I'm getting a little of everything here," Borlaug declared, "but it's not taking me anywhere. I'm unhappy here! I belong somewhere else."

Dr. Hovde looked at his record, remembering the determination he had seen. "Well now, Borlaug. If I agree to grant you a transfer, what other college would you select?"

He showed no hesitation. "The College of Agriculture—I'd major in forestry."

Dr. Hovde knew the growing appeal of forestry among young students. It meant a life in the open, and many of them adopted the garb of rugged foresters while still at the university—check shirts and hobnail boots. They let their beards grow and assumed the manner of tough lumberjacks; it was almost a cult. But Hovde decided that glamour was not the reason for Borlaug's selecting the school. The young man was too serious about what he wanted to achieve.

Dr. Hovde finally nodded his agreement. He could enter the College of Agriculture after completing the quarter. Borlaug left the office walking on air. The last days of the quarter were his happiest since leaving high school. He now had an objective, a future. He forgot his money problems for the moment and unconsciously adopted Champlin's belief that doors would open when most needed.

He went home for the Christmas break full of confidence. He and Upton hitched their way to the Iowa border, two young men looking forward to the holidays and the start of a new year.

When they returned to the rooming house in Minneapolis in January, they were met by a downcast George Champlin. The coffee shop had gone bankrupt, another victim of the Depression. Their three meals a day had vanished and at once they were beset with financial problems. Borlaug had thirty-five dollars—but twenty-five was earmarked for tuition. Upton was in similar straits. There was rent, heat, books, as well as transportation. Champlin said they could take another student into the room and suggested Scott Pauley. In addition, he offered to bear a greater share of the rent, bringing their share down to a dollar a week. He also offered to drive them downtown in search of restaurant jobs that would pay them in meals.

They had some success, but the work was intermittent and uncertain. On evenings when the restaurants had no clients they would be turned away hungry. Borlaug made a few nickels and dimes in tips by parking cars when concerts were held, but even this work became harder to get as the number of students competing for such jobs increased.

Just as their situation was at its blackest, Scott Pauley came bounding up the stairs to the room, eyes shining, breathless and laughing. "Work and food—for all of us!"

He was waiting on tables, he said, at a sorority house, serving meals to the daughters of the privileged rich! And there were jobs for two more students, named Borlaug and Upton, if they hurried!

That evening Pauley took them to meet the housemother, Mrs. Nichols. It was arranged that they should start immediately, and for the next two years Pauley and Borlaug served three meals a day, were treated with great consideration, and were given all they could eat. Mrs. Nichols became a second mother to the boys, and Borlaug remembers her with fondness and gratitude.

Hot on the heels of this success came the first help from the National Youth Administration—a relief project started by President Franklin D. Roosevelt. The NYA provided funds for university staffs to allot to students who were in need of work, and the money filtered through to Minnesota early in 1934. Borlaug was called in one day and interviewed by Dr. Mickel of the College of Agriculture. There was a job with the department of entomology—pinning up insects, tidying up after the professors and graduate students, and being generally helpful. It would pay twenty cents an hour, fifteen hours a week, and was his if he wanted it. It was munificence! Three dollars a week!

With meals at the sorority house and this money, he could manage well.

It forced him into a rigid time schedule, however, which he had not taken into consideration. Waiting on tables took more than thirty hours each week, and another fifteen hours for NYA made a total of forty-five working hours that had to be fitted into a full academic schedule: classes, lectures, reading, travel to and from the farm campus, and wrestling.

Time became a straitjacket. He was out of bed by five-thirty each morning to put in an hour of study before going to the sorority house to serve breakfast at seven. Ninety minutes later he had to be across town, five miles away, attending his first class at eight-thirty. His morning schedule was arranged so that he could travel back half an hour before noon to wait on tables at lunch. Classes started again at one-thirty and went through to late afternoon, after which he went, five days a week, to meet Upton at the gymnasium for ninety minutes of wrestling and training. On those nights they would have to shower and dress quickly so they could reach the sorority house by quarter to seven to serve the evening meal. On weekends, and on evenings when he had no classes or lectures, he worked at his NYA job in the department of entomology. It was too much. He cut down his sleeping hours, rising at five, sometimes half-past four, in an effort to stretch the busy day.

He saw little of Margaret Gibson; they would meet briefly on the campus and spend whatever time they could together on Sundays. The separation was difficult for both of them, and Margaret was concerned that the work schedule would break Borlaug's health.

In trying to make use of every hour, Borlaug and Upton found wrestling to be more important than mere recreation —despite its demand for physical effort and vigor. They had experienced a highly successful season, and their intercollegiate successes had attracted a growing band of devotees. One afternoon they held a meeting, all seated around the wrestling mat, to discuss their need for disciplined training and better coaching. The part-time Minnesota coach, Blaine McKusick, a lawyer, was retiring. Borlaug said he and Upton knew of only one man who could put Minnesota at the top of university wrestling: Dave Bartelma of Cresco. They wrote to him for his views, and he said he would be happy to move to Minneapolis provided he could also study for a doctorate in psychology. The students then formed an action committee, presenting their views to several faculty members for consideration, and the campaign took shape. There were no great difficulties. By the time the summer came the issue was settled; Bartelma was invited to take charge of the wrestling team the following September.

Borlaug went home for summer vacation, thumbing his way to Saude. He longed for the long, hot days in the fields with his father, fishing in the Turkey River, yarning with old Nels in the shade by the vegetable patch—seeing all the old places and faces, dancing on Saturday nights over at Spillville or Protivin. He was doomed to disappointment.

The countryside lay under a pall of gloom. Both sides of the family—the Vaalas and the Borlaugs—were apprehensive. All about them friends and neighbors struggled to avoid foreclosure. Prices had collapsed in an appalling paradox. There were gluts of produce—not from growing too much but from too few people having money to buy it.

He soon learned that families he'd known from childhood had been defeated and driven from their properties, which had been taken over by banks and sold for a fraction of their worth. Men, women, and children had been left sitting with their belongings in the dust of the road, outside the gate of a farm they'd worked all their lives—homeless, beaten, nowhere to go. He witnessed the forced sales held under shotguns—sheriff's officers ringed by menacing crowds of countryfolk who wielded pitchforks and rake handles. It was a terrible, angry summer.

Then, too, his grandfather was not well. Old Nels took to his bed, grumbling at the discomfort in his stomach, blaming his wife for cooking that was not as good as it used to be, even though he knew it was not the truth. And to make matters worse, there was a terrible drought—the worst the oldest farmers had ever seen.

For Norman Borlaug the summer of 1934 was a time of shadow.

That September, when the academic year opened, Dave Bartelma blew into Minneapolis like a cool, fresh breeze, stirring activity all about him. At once he had the wrestling team organized, weights classified, and training schedules worked out. Bartelma was very, very firm. He looked at his charges as he had done when they were high school students at Cresco, shaking his head in doubt.

"You've been eating too well, Borlaug. Too fat. You'll roll across the mat instead of fighting! Have to do something about that. Get you down to 145 pounds—and keep you there!" He prodded stomach muscles, felt biceps, and tested fingers. All the wrestlers had to be strengthened and toughened, or they might just as well take up chess.

Quickly Bartelma sketched his plans. The state would have a network of wrestling teams, such as had been founded in Iowa and Oklahoma, and they would then be expanded to interstate competition. In this diadem of wrestling the University of Minnesota team would be a shining jewel.

It would take years to build up such a team, but even in the first weeks, each new meet attracted bigger crowds at the university's fieldhouse. Margaret Gibson came to watch and took pride in Borlaug's victories. As the season neared its end, the tension mounted. The final tournament would be with the championship Illinois team in Minneapolis. Bartelma again told Borlaug he would have to get his weight down—by ten pounds to 145 pounds.

"Your work and study schedule will make it difficult. You'll have to starve, spend hours in the sweatbox, and take very little water for at least three or four days. Not enough to weaken you—but enough to get you rangy and fit to fight." Bartelma's instructions were not to be disobeyed.

One day after another dragged by without food. He waited on tables but did not eat, or even drink coffee. After the third day he stopped drinking water to reduce his weight still further. He went for a walk one evening with Margaret and Scott Pauley, and Margaret told Pauley later that she thought he was extremely irritable and tense.

On the afternoon of the Illinois tournament Borlaug went into a sweatbox for an hour while Bartelma joked with him, joshing him for sitting idle while other men were training hard on the mat. It was his fifth day without a meal. When he left the box he went, naked, to the weighing scales and stood there looking at the measurement. Oh, God, he thought, still a pound overweight. One of the other wrestlers came to look at the scales and put his hand on Borlaug's arm.

"Let's see how you've done, Norman."

Borlaug's outburst was explosive, uncontrolled. His rage was motivating him without thought, without hesitation. He wheeled on the young man with whom he had always been friendly.

"Take your filthy hands off me, or I'll break your arm!"

He was about to smash a fist into the startled face when Bartelma stepped in between them. "Take it easy, take it easy," Bartelma said as he led him toward the showers. Standing under the cold water, Borlaug discovered his flash of anger had come and gone, like a lightning strike, and shame flooded through him. He looked at Bartelma. "I'm sorry," he said. "I behaved like an animal." Bartelma told him it was understandable. "Don't worry. I'll explain it to your friend."

Borlaug went back into the sweatbox, lost the last pound, and won his match that night. But there was no pleasure from the victory, even though Bartelma had arranged a buffet afterward. Borlaug wolfed down the food but would not stay for the singing because he was still troubled by his outburst of the afternoon. Walking home with him that night, Margaret listened to the story, aware that it concerned him so much because the act was alien to his character. His friend had merely laid a hand on his arm, he said, and "I turned on him, snarling and threatening, like an animal." He would apologize, that was easy to do, but the memory of how he had behaved stayed with him. He had been violent, raging, like those people he'd seen in the milk strike riot.

"I think I've learned a primal rule of nature," he told Margaret. "You see it wasn't me at all. It was primitive, rudimentary. I can't explain how hungry I was. I was starving, and I found out that a hungry man is worse than a hungry beast."

Three

In his sophomore year Borlaug found new challenge in his studies. Forestry became a love, a new world that went far beyond the enjoyment of a life outdoors and the wearing of check shirts, beard, and hobnail boots. It was a desire to become a scientist, to understand the complex ecology and evolution of the forest. He studied hard, for he knew that he had much to learn. His first important examinations would be held in the spring, and it was essential that he pass them.

The examinations were to coincide with family tragedy. Late in April 1935, halfway through the testing, Borlaug was called from the classroom and asked to report to the office of Dr. E. G. Cheney, his professor in forestry. His father had telephoned from Iowa and he was to call back, collect.

His father's voice was sad; controlled, but quieter than usual.

"I don't know how to say this, son. But—he's gone. My father—your grandfather."

Borlaug was at loss for words. They had known, of course, that it would come. When he had left home at

Christmas, he had felt it might be a last good-by. The old man was bedridden, weakened by cancer of the stomach, but alert and clear in his mind. The tumor was invincible. It was inevitable, but it did not stem his father's grief. It was still a shock, a blow.

"Look, Dad. I'll come home straight away. I can take my exams again next quarter—"

Suddenly his father's voice was clear and firm. "You will not do that at all, Norman. It would go against his wishes, my wishes, and your mother's. He was pleased how well you're doing. He wouldn't want you to come home and interrupt your work. Later, perhaps. But you stay there now and help by doing your best."

Norman Borlaug felt shattered when he put down the phone. His grandfather was dead—gone.

He would be buried, of course, in the grounds of the church his kinfolk had built at Saude. He would join his mother and father, Solvei and Ole, whose gravestone bore the legend that had become a family text: "In Thee O Lord Do I Put My Trust." He would rest near his wife's family, the Swenumsons.

Norman struggled through the rest of his examinations and made an extra effort to pass them, for he knew that was what his grandfather would have wanted. Nels Borlaug would not have tolerated the nonsense of hindering education for his burial. "Education, Norm, puts vital power into a man," he had said during their last conversation. "Fill your head now if you want to fill your belly later on." And after a pause the words of advice he had given before: "Just don't stop asking why things are the way they are, Norm. One day someone'll tell you the answers."

Margaret was moved by Norman's deep grief: "Oh, Norman, love; I'm so sorry." She was sympathetic, and her arms gave him solace. He had often talked to her of old Nels, and she wished now she had known his grandfather so that she could be closer to him in his sorrow.

He did not go home that summer because he had accepted a job near Rochester, Minnesota, to earn money for the coming term. The previous Christmas, when he and Erwin Upton were hitchhiking back from Iowa, they were picked up by a man named Jack Chrysler, a field manager for the Reed Murdock Canning Company in Rochester. When he heard of their financial difficulties, he offered them a summer job.

Early in July they started work in the company's fields near the little town of Spring Valley, Minnesota, pulling pea vines and loading them onto trucks for dispatch to the canning plant. They were paid fifteen cents an hour for a twelve-hour day— from six in the morning to seven in the evening, with an hour for lunch. They made $1.80 a day, but out of that they each paid a dollar a day for a small room in Spring Valley.

When the peas had been picked, they were shifted to sweet-corn harvesting. One day Jack Chrysler came and said to Borlaug: "I need a timekeeper for this work. Will you do it?"

It raised his pay to eighteen cents an hour; but it meant a longer day. The crews worked in ten-hour shifts, and he had to check both shifts. At the height of the harvest, when the corn was ripe and had to be rushed to the factory for canning, he worked a twenty-hour day, with only two hours' rest between shifts. He received $3.60 a day.

When the harvesting was done, Borlaug and Upton rested for a full twenty-four hours, then took a bus. They were too tired to walk the roads thumbing rides back to school. The work had been exhausting—but they had their tuition fees for the coming term.

For Borlaug the fall passed quickly. He and Margaret began to see each other regularly. They made no real decision about their future and what they would do, but the understanding was that one day they would marry—when the time was right. Then, in December, Margaret decided to end her university studies. She could not tolerate the scrimping and scraping any longer, she told Norman. Her brother Bill, editor of an alumni magazine, learned that the Colwell Press, which printed his publication, was looking for a young woman to serve as a junior proofreader. It wasn't wonderful money to start, but it would mean escape from the girls' boarding house and an end to the constant struggle to earn enough to eat. "Maybe I'm not as determined as you, Norman, but I'm sick of going to bed hungry at night. I want a few of the little things a woman needs in her life."

It also meant, she said, that she would be able to help him when the going got rough, but she knew he would not accept money from her. He was already planning ahead for a summer job to pay his way. He would not spend another summer as he had done in Spring Valley, though. He would look for something in forestry that would further his work and also pay better. He decided to write to various regional forestry offices—fifty-five addresses in all—and he asked Margaret's help with the typing. She borrowed a machine and spent evenings pounding out letters.

Borlaug was not overly optimistic. "I don't suppose there's much chance of anyone hiring a second-year student as a junior forester, but we can try."

In his circumstances even the investment of fifty-five postage stamps was a gamble, but the letters went into the mailbox with Margaret kissing each one good luck. They awaited a reply with a sense of adventure and excitement.

Meanwhile, Borlaug obtained another campus job through the National Youth Administration. It involved two hours, three mornings each week, of cleaning cages and feeding experimental animals at the veterinary department. It meant rising at three-thirty in the morning and riding the "owl" streetcar that rattled its way to the farm campus at four. One bitterly cold January morning he left his topcoat on a seat outside the heated animal house, and when he came out it was gone. He was left with only a thin buckskin jacket and a sleeveless wool sweater. He joked about his carelessness, but Margaret worried that he might catch cold and become ill.

"I have some money, Norman. Take it and get yourself another coat—anything, so long as it keeps you warm. You can pay me back later."

But she could not persuade him. It was not right to him and nothing altered that sense of right. He also rejected her plea that he write his parents and borrow money for another coat. His resolution did not weaken even when the temperature fell to twenty or more degrees below zero. It was a bitter, icy winter; yet Margaret Gibson would see the young man from Iowa come down the street on a freezing night, hands stuffed into trouser pockets, shoulders hunched inside the thin jacket, face alight with the warmth of his smile and delighted to see her, and he would talk animatedly about the activities of the day. He never seemed to tire. All winter long Margaret was amazed at his resilience and motivation. It set her wondering at first, then marveling: he did not falter, or have second thoughts about the work he had to perform, where he was going, or what he would do.

The dozens of letters they had sent around the United States for a job in forestry brought only a single reply. The Northeastern Forest Service at New Haven, Connecticut, sent a letter offering him temporary employment as a junior forester as soon as his May examinations were completed.

He would be paid fifteen dollars a week and had to find his own transportation to New Haven, where he would be interviewed and given his assignment.

By now he was an expert at hitchhiking. Immediately after exams he thumbed his way to New York and then on to New Haven. He was given a post at the Hopkins experimental forest station in the Berkshire Hills, near Williamstown, Massachusetts, a setting of beautiful forests and woodlands in lovely, rolling hill country. The station was on an old estate, once owned by a railway magnate, that had passed into government possession. It was supervised by Boch Starr, the forest engineer, who lived with his wife, Emma, and their two children, Caroline and Bud, in one of the carriage houses. The old mansion still stood but was too rambling for comfort, so Borlaug was allocated a small cottage and he ate all his meals with the Starr family.

It was a wonderful summer. His work was engrossing. He was responsible for laying out experimental plots of young trees, for clearing areas of dead forest, and for making boundary surveys and studying the diseases that afflicted the trees of the region. He was only twenty-two but he was put in charge of a crew of four young men from the slums of Boston, his first experience with leadership and command. The men were recruited from the Civilian Conservation Corps, one of President Roosevelt's programs for relief of the poor and the unemployed. They were given barracks-type quarters in the old buildings and had to cook and wash for themselves. But they had work—even if their pay was less than five dollars a week. Borlaug mixed with them and asked about their lives before they were recruited to the CCC. There were signs of resentment and bitterness as they talked of mass unemployment and lack of privilege, but Borlaug understood their feelings and had no difficulty winning their respect. He asked no member of his crew to do work he was not ready to do himself, and he worked harder than any of his men.

At the end of the summer Boch Starr asked Borlaug to stay on and help him until the end of the year. Borlaug accepted, arguing to himself that while he would forgo a quarter of university study, the practical experience would be invaluable. And, of course, he could use the money. But it was more than that. He wanted to work among the trees —it had appeal, it struck a chord in his nature. And though he missed Margaret, he enjoyed every day he spent in the forests of Williamstown.

Borlaug returned to Minneapolis early in January 1937, with the assurance of more work in forestry the next summer. With the wages he earned at Williamstown he no longer had to rely on the NYA job. He also had enough money to pay for his own meals and could quit waiting on tables at the sorority house. He concentrated on serious study and continued his wrestling with Bartelma, entering a series of tournaments.

Meanwhile, Margaret was doing well in her job as assistant proofreader, and life was easier for them than it had been in any previous semester. They spent more time together and talked of their future; but still the final decision was deferred by the uncertainty of the times. He was determined to secure a future before he would ask her to marry him. That seemed quite sensible to Margaret.

The winter months flew by and in March the letter that Borlaug had been expecting from the National Forest Service came. He was to report in May to the forest service depot at McCall, Idaho, for special training. He packed his bag and said good-by to Margaret. There was no indication of what was to come.

In McCall, Borlaug was one of fifteen young foresters

being trained to man a network of observation and fire-fighting posts located strategically throughout the vast primitive forest areas of the state. The junior foresters were under the supervision of veteran rangers, and their training was very practical. They were taught the principles and techniques of fighting forest fires and of establishing and maintaining telephone and radio links in the remote areas for which they were responsible. They were also tested carefully for reliability, capability, and self-reliance.

Borlaug was among the best performers in all areas, and he was called into the senior ranger's office one day to face a small panel of examiners. At the close of the session the spokesman said: "You'll go to one of our key posts, Borlaug. It's a vital link in the fire protection network and we'll be depending on you. It is also the most isolated post in the U.S.A."

He was to leave the following week for Cold Mountain, in the Middle Fork area of the Salmon River district, one of the biggest and most remote wilderness areas in the United States. He would be there on his own, with sole responsibility, for the entire summer. It was a job that was to test the full range of his resources.

A few days later the small convoy of forest service pick-up trucks rolled out at dawn, carrying the squad of apprentice foresters to their scattered posts. They would spend the summer watching and scouting, reporting the blue smoke signs of possible devastation. They would likely be the first men on the scene of such outbreaks, guiding in the mobile teams that would come later.

Borlaug was dropped off at midmorning at the Big Creek Ranger Station near the end of the road to Cold Mountain. The post where he would serve was forty-five miles away across trackless, rugged country. The pack-horse driver, a taciturn man named Jack, was irritably anxious to start and quickly strapped Borlaug's duffle-bag on the back of a horse. Moments later they were riding into the trees, climbing among shattered rocks and giant pines to the long, sloping escarpment that was the objective of their first day.

They spent two nights in a small tent in sleeping bags, protected from the soaking mountain dews; each day they climbed higher and higher, the packman picking his way unerringly through the jumbled, rock-littered landscape. On the third morning he pointed a finger toward the mass of a great mountain shoulder looming against white clouds.

"Up there aways, sonny. Should make it by sundown—thereabouts."

The packman stayed the night in the cabin on Cold Mountain. Soon after daylight he ate breakfast with Borlaug, and with a laconic "See ya in about six weeks when I bring supplies," he was away down the slopes with his horses, slipping and stumbling on the broken surface.

Borlaug was alone. The immensity of the wilderness closed around him. He had been trained for the work—to maintain communications, watch for fires, and study trees and living things—but he had not been prepared for the solitude and challenge of utter isolation.

He stood on the flat slab of rock that would support his home for the coming months: a one-room lookout shelter perched atop a steel tower 8,600 feet above the great Salmon River, and a one-room log cabin at its foot. He looked out across a huge panorama of twisted, buckled landscape, with the Rocky Mountains rising and fading into the haze of the far blue distance.

This was the primitive region of the Idaho National Forest, high above the river junction known as Middle Fork, and the terrain fell away from him on all sides, except behind his cabin. There were massive forests as far as the eye could see, and below the mountain slopes the land folded into deep canyons and was broken and crossed by ravines.

The sound of whispering pines seemed to come with greater intensity as he lay in the dark in his sleeping bag thinking of the problems he would have to face with the coming day.

The location of his cabin had not been chosen for comfort or convenience. It was primarily a spotting post, exposed and situated for surveying as wide an area of the wilderness as possible. The water supply was a problem. The nearest pool was more than a mile down the mountain, a clamber over rough, broken surfaces that took him twenty minutes going down and more than an hour coming back. He searched higher up the mountain and found a spring that he diverted by constructing a small channel leading to a deep rock hole. This created a reservoir a mere five minutes' walk from his cabin and provided enough water to meet his daily needs.

In the cabin he had a table, two chairs, and a bench on which he mounted the radio and telephone equipment. From here each evening he made his sole contact with the outside world—a telephone hookup with ranger headquarters according to a prearranged schedule. Occasionally the spotter aircraft flew overhead and he had contact with it over the radio, but those communications were infrequent and usually marred by static and interference.

Borlaug was never lonely and never bored. He ate, washed, and worked alone, but his days were full of fascination. Outside the door of his cabin the great trees of the forest were engaged in their relentless struggle for existence, a constant battle for food, living space, air, and light. Here in the forest he witnessed an ecology that pitted one life against another: the climbing, choking vines that used the trees to reach the light and killed their hosts in the process;

the fungi that sucked away the sap; the insects that preyed on trees and on other insects; the birds and animals of astonishing variety and habits that showed remarkable ingenuity in trapping their prey. He came to believe the evidence before him. The so-called balance of nature was a myth. Thus the dominance of type and variety depended not on biophysical balance but on circumstance and adaptability—survival of the fittest. One afternoon he stood in the lookout tower and watched a small lightning storm start eighteen forest fires. He saw nature as a constant state of dynamic change, in which environmental pressures came from droughts, fires, floods, frosts, insect diseases, and invasion of the habitat by other species. It was not an equilibrium but nature out of balance that created evolution—and it was an insight that was to have profound effect on his views.

He never tired of observing his surroundings. High up behind his cabin reared the bare mountain peak with its cap of constant snow, and below that snowline were masses of fir and spruce, merging lower down the mountain with other types of fir. Lower down still were lodge pole pines, which in turn gave way to the more prevalent ponderosa, or yellow pine. There were dry areas in some of the canyons, where sagebrush grew, and there were valleys fed with streams, where the swampland was vivid green with sedge and reed grasses and where he sometimes wandered close to the grazing elk, his presence causing them no more fear than if they were domestic cattle.

There was also daily peril as he patrolled his region, and caution was imperative. A careless step, a fall, a broken leg, and he could lie for days without help at the mercy of the elements and roaming animals—cougars, mountain lions, bears. In the dry ravines were rattlesnakes. In the moist areas lurked a far more insidious danger—the skin-burrowing ticks, which were the agents of Rocky Mountain fever.

Each evening he stripped and searched his body and clothing for these revolting parasites.

After a few weeks on the mountain he had a great yearning for fresh bread. He built a fireplace out of clay and fashioned a field oven from a biscuit tin. Then he mixed the flour and kneaded the dough as he had been taught by his grandmother. But it ended in frustration. At that altitude bread would not rise, and he was forced back on the old sourdough preparation known to prospectors. His diet was simple, mostly canned and dry foods, but he learned to add to his menu. He found streams where the trout were so hungry they could be caught with an improvised fishing rod—a stick, a piece of string, a bent safety pin, and grasshoppers for bait. He caught an occasional salmon, and sometimes he shot a grouse; but though he saw many game animals, mostly elk, deer, and grizzly bear, he left them alone. He had no refrigeration. It seemed a waste to kill for a single meal or to satisfy a whim when he had other food.

The packman came again in mid-July, leading his panting horses up the slope to the cabin. He brought fresh supplies: batteries for the radio, newspapers, and mail. He was the only human being Borlaug had seen in six weeks, but it was a brief interlude. When man and horses went stumbling back down the trail next morning, Borlaug sat down and reread Margaret's letters carefully. Her main news was that she had changed her job. She was now a full-fledged proofreader with a firm called Independent Press in Minneapolis, and she had rented a small apartment not far from the campus. She was full of chatty news about her job, mutual friends, campus activities. She was careful not to mention how much she missed him, but Borlaug sensed it and for the first time felt a pang of loneliness.

The newspaper items seemed echoes of another world, as indeed they were. But though he was isolated from world events, he felt in some way closer to the mainstream of life than ever before. The solitude gave him time for reflection, and he observed his surroundings with new insights.

That evening he walked out to the edge of the flat rock to watch the flaming cascades of color as the world turned away from the sun. The line of Rocky Mountains was silhouetted in a splash of light that flooded into the clouds above his head. Then the clouds piled higher, turning black, and he knew what was coming. Soon streams of water were running from his roof, splashing onto the rock, and the blackness was split by streaks of lightning. It was a sudden, savage storm but, to his forming thoughts, part of the order of things. In the movement of the earth, the pattern of the stars, the constant growth and renewal of the forests, he saw an order imposed by a power greater than man. It gave him his basis for faith, not a formal religion but a belief in something all-powerful that strengthened and sustained him for the rest of his life.

The summer brought other storms—and fires.

On one occasion, when lightning brought down vital telephone wires, he went out to make repairs and was trapped by a storm for two days on a mountainside. Without food or shelter he had crouched under a makeshift lean-to of tree boughs, waiting for the raging torrent to subside before he could make his way down the treacherous slope. Another time he spotted a fire twenty miles down the valley and hurried to help a team of mobile fire fighters being dispatched to control the dangerous outbreak. He saw the fearsome nature of forest fires, their flames leaping from treetop to treetop in great spurts, and was filled with admiration for the courage and the skill of the men who faced these dangers.

Summer turned quickly into fall, and one day in mid-September the packman came to take him and the equipment from Cold Mountain. The first snows were already falling. His wilderness looked like a Christmas card, but its beauty was the cold beauty of death—for a season. He retraced the forty-five miles he had come those months before as a tenderfoot forester. Now he had matured. He had become a man who could live with himself in the wilderness and survive; a man of greater physical strength, inner resources, and confidence born of self-knowledge.

Borlaug's performance immediately won him an offer of a full-time job as forester with the Idaho National Forest Authority. He was elated, and accepted at once. He was to start in January 1938, after completing his degree.

He traveled by car to Minneapolis, surprising Margaret as she left the printing office one evening. He told her of the offer of steady work and then blurted: "There's no reason to wait any more. Will you marry me?"

They scheduled the event for the following Friday, at seven in the evening. It was to be held at her brother Bill's, who had married and settled down in Minneapolis. Borlaug telephoned his father and mother. They could not leave the farms at harvest time, but his two sisters, Palma and Charlotte, said they would take a train and come.

Margaret finished work at noon on Friday and treated herself to the luxury of a hairdresser. Her brother's wife, Lois, came to meet her and together they shopped for last-minute items to complete her modest trousseau—a handbag, much more expensive than she could afford, and a pair of new shoes.

At her brother Bill's she relaxed in a warm bath before dressing for her wedding, and by six she was ready, glowing and waiting. At seven the local Lutheran minister arrived, followed by Borlaug smiling and neat in a new gray suit. Their long-time friend Scott Pauley was best man. There in the sitting room they gave their vows in quiet voices and kissed each other tenderly.

There was to be no wedding trip; they had neither time

nor money. After a brief reception they made their way to Margaret's apartment, the first home of their married lives, to spend their weekend honeymoon. It was a modest one-room flat, with a table and chairs, a day bed, a stove and sink in one corner, and the bathroom down the corridor, shared with other tenants. But it was clean and neat, and Margaret had made it cozy and attractive, with bright curtains, a few pieces of good crockery, and vases of fall flowers.

Monday was a working day, like all weekdays for the rest of that year, and they quickly established a pattern. Margaret would rise at six, and cook her husband breakfast—she knew he would have no more than a cup of coffee the rest of the day. Then they would walk together to the streetcar station, where they would go their separate ways for twelve hours or more. Margaret would ride downtown to her job at the printing plant, and Borlaug would sit with his nose in a book headed for the agriculture campus. They had to be content with brief exchanges in the morning rush and at night when he came home tired and hungry—and invariably late. Margaret quickly came to terms with the dominance of work and study over all else. It was essential that Norman complete his bachelor's degree before the end of the year, and she never begrudged the long hours he devoted to that pursuit. Instead she took pride in his motivation and encouraged him in his work.

During that fall term, Borlaug met a man who was to influence the course of his life. One afternoon in the forestry pathology laboratory he looked up from his microscope and saw a square-built man in his late forties watching him over the glowing bowl of a pipe. Borlaug had been scrutinizing wood samples, trying to determine which kinds of fungi left which kinds of stains. The visitor did not ask his name but immediately began questioning him—not about the problems of fungi, but about the wood samples themselves: What types of trees were they? What were their anatomical and chemical structures? Why did one type accept a stain when other types did not? What made them different? The inquisition was at Ph.D. level and Borlaug bridled at the way the questions were thrown at him, but he answered with a logic that seemed to satisfy the intruder.

His teacher, Dr. Clyde Christensen, came into the room and soon the visitor left, with a handshake and a smile. Dr. Christensen told Borlaug he had just been quizzed by Dr. E. C. Stakman, head of the department of plant pathology and one of the most respected scientists in the field.

Not for some years did Borlaug learn that Stakman's appearance and tactics were neither haphazard nor accidental. They were a device to stimulate reaction and to measure his response to challenge. Stakman had seen Borlaug in a wrestling contest and had been struck by his courage and tenacity against a stronger opponent. The professor had thought that if Borlaug possessed similar tenacity in his intellectual pursuits, he was good potential for scientific research.

Weeks later Borlaug saw Stakman's name on a bulletin board. He was giving a special lecture on the nature of rust diseases in cereal crops. Borlaug's own work on trees had involved him in a study of rust—a parasitic fungus that attacked a wide variety of plants, sapping them of their nutrient juices. The disease was characterized by reddish-brown spots on the leaves or stems of the host plant and could wipe out huge crops in a matter of days. What made it so insidious was its ability to adapt, foiling each new defense against it with another mutation. The lecture could be helpful to Borlaug's courses and he decided to attend.

Dr. Elvin Charles Stakman had pioneered in the research of rust diseases in crops before World War I. He had discovered entirely new races of these organisms, and it was he who had shown how their capacity to adapt could threaten the food crops of the whole world. He was an international authority, but as he stood on the platform in the crowded auditorium that night he was more than that: he was a magnetic and compelling teacher. His style, his sincerity, the intensity of his delivery made his words ring in Borlaug's ears.

The lecture was more than just a summary of recent developments in research; it was a magnificent philosophical discourse on the phenomenon of rapid evolution. The speed with which rust spores could change their character suggested a response to a form of genetic intelligence. Science could breed new defenses into cereal plants by crossing the pollen of one plant with the stigma of another. But victory was always temporary. "Rust is a shifty, changing, constantly evolving enemy," said Stakman. "We can never lower our guard. These floating fragments of beautifully engineered genetic matter are an artwork of nature—but they are the enemies of human appetite. Rust diseases are the relentless, voracious destroyers of man's food, and we must fight them by all means open to science."

But there was another threat to man's food supply, warned Stakman, and that was man himself. The challenge of human numbers against available resources was not new. Hunger had visited the table of every civilization, and mankind had failed to solve this fundamental problem. Now a new element, science, held out hope. With a combined program of population control and rust disease eradication, science might wipe out the specter of human starvation.

"Do not deceive yourselves—ever—that scientific approach is omnipotent," cautioned Stakman. "It will make its mistakes; but it will go further than has ever been possible to eradicate the miseries of hunger and starvation from this earth."

It was a moving lecture, and as Borlaug walked home

in the cold night air he mulled over Stakman's message. Science, diseases, food, population, world hunger, all were interwoven.

He related the bare details of the lecture while Margaret cooked supper, quoting phrases from his notebook. As he spoke, the intensity of his voice, its tone, drew her attention. She studied him carefully. This was her husband, stimulated into great mental excitement, and she wondered at the chord that had been struck in his mind. Suddenly he stopped talking, and his expression changed. Then he said: "One day I would like to go back and study under that man if it is ever possible."

They were silent for a moment. They both knew he would soon go back to Idaho and become a forester, earning the money to give her a better home, a better life. She said to him, consolingly: "Perhaps you will one day. Who knows."

It happened sooner and more unexpectedly than either dreamed. Two weeks later a letter came from Idaho, and he stood in the middle of the room reading it. There was to be no employment in the forests of Idaho. No job in the new year. Budget cuts had forced reduction of staff and this was deeply regrettable. If he cared to reapply later there might be casual employment during the fire season. It was the first setback of their marriage.

It was two weeks before Christmas. Norman was in his last days at college and confident of his degree. Now their plans and hopes were dashed. The savings from his Cold Mountain job had dwindled. They faced a rent of thirty dollars a month and the prospect of having to pay all expenses and bills from Margaret's wages. He felt small and useless. Where would he find a job? And what sort of job? Margaret saw his growing despondency and tried to cheer him up. It was not a great problem, she said. There were other jobs; it was a matter of time, nothing more. All he had to do was put that time to useful advantage. She was confident that he was postgraduate material and that it was what he should do. Then she remembered his words on the night of the lecture.

"That man you wanted to study under—Stakman? Why not now? You'll have the time. We'll manage for money, somehow."

He could try a term of studies she said. After that he could go back and work in the forests for the spring and summer and save enough to get through the next fall and winter sessions—if he wanted to do that. But Borlaug also wanted to provide for his wife.

"And what of you, Margaret? What would that mean for you?"

She laughed away his doubts.

"Oh come now, Norman. I'll go on working for a while —that's all. I enjoy it, really. You know that. Sitting there all day reading all kinds of interesting things—and I get paid for it."

They could just manage on her wages. But his sensitivity made it more than a question of managing, or going without things. It worried him that he was not being productive.

"Go and do postgraduate work," she urged. "That will give you an objective. We can talk about it later when you're clearer on what you want to do." Her persuasion restored his confidence.

Next morning he walked boldly through Dr. Stakman's outer office and knocked on his door. Stakman believed that professors should be available to students who needed help and guidance. His office was always accessible and the knock brought a quick response. He was not surprised to see Borlaug. It was typical of what he knew of the young forestry student that he should walk in and ask for help without ceremony.

Stakman heard him out, fingering an unlit cigar, looking at the young man under his arched brows. He sorrowfully shook his head, in some doubt.

"Fill in a month or two—between jobs? That's a pretty poor reason for asking for postgraduate study, Borlaug! You should know better than to take a view like that. It's not like a novel you can pick up and put down. You'll have to be a bit more serious about it than that, my boy."

And what did he want to study? Did he know that?

Of course, said Borlaug. He wanted to continue with forest pathology. Again Stakman shook his head—almost in resignation.

"Look here, Borlaug, for heaven's sake get your feet on the ground. If you take advanced education in forestry—bingo!—you're stuck with forestry. You will have great trouble transferring or switching to something else. Use a little common sense. If you broaden your base, you can still qualify for forestry, but you can also do other things. Take plant pathology and that will keep you out of narrow specialization. You'll get a wide sweep of agronomy—the genetics, soils, and plant diseases—and you'll be equipped for much more than trees."

He told Borlaug he could enroll for postgraduate work, and that he would help him find a paid assistantship for the coming year. The amount would be small, but it would help.

"Put away this idea of taking higher education in dribs and drabs, Borlaug! Aim for your master's degree. Look beyond that for a doctorate and I'll go to bat for you. Play about with your education and you'll have to find help somewhere else."

So Borlaug enrolled for a year of postgraduate work under Stakman. He and Margaret enjoyed the months of 1938, despite the struggle to make ends meet. They did not have much, but then few people they knew had much. They picked up a few bits of furniture from time to time but spent none of their income on entertainment. At the end of the year they discussed future education, and after more consultation with Dr. Stakman Borlaug decided he would continue until he earned his master's degree. Stakman said he would try to get him a job that would allow him to study and to make a contribution to their living expenses at the same time.

It was not long before Stakman found a research assistantship in the department of plant pathology. It meant that Borlaug would not have to pay tuition and would receive a small salary. It also advanced his studies, and by the end of the year cereal pathology was opening new doors to him.

In the new year of 1940, after three years of married life, Margaret and Norman were able to move out of their single room. Borlaug was made an instructor and they were offered a university-owned apartment, close to the main campus. It had four rooms, with every convenience, and the rent was reasonable.

That summer was the best that the young couple shared together. Borlaug received his master's degree; his mother and father were happy and proud of their son, the first graduate in the Borlaug family. Old Nels had had visions of this day, and he was missed again. Margaret was happy for her husband, but she knew he faced a tough struggle for his Ph.D.

Even before he had completed the examinations for his master's, he and Margaret had decided he should go on to get his doctorate. They were not sure how he would use it, whether his career would be in academic, commercial, or public life; but they knew it would open doors that had been closed before.

He was still in the process of writing his thesis in October, 1941 when Dr. Stakman asked him to call at his office to talk to his former forestry tutor, Frank Kaufert. Borlaug could not imagine what they wanted. Was there an opening in forestry they wished to discuss?

Kaufert had taken his doctorate under Stakman and had spent the past three years at E. I. Du Pont de Nemours & Company in Wilmington, Delaware, heading a biochemical laboratory group. He had come back to Minnesota as an associate professor of forestry and was to be dean of the department.

He greeted Borlaug warmly and launched into his proposal. The Du Pont people had asked him to recommend his successor and he wanted to propose Borlaug for the post. He would have direction of the laboratory, within the prescribed program, and would receive a salary of $2,800 a year. The sum sounded enormous to Borlaug. Stakman said he would back the recommendation. Borlaug could easily complete his thesis in a month or so and look forward to his doctorate the following year.

It was a good offer. The Wilmington laboratory was beautifully equipped and was one of the few commercial biological research units outside the drug industry. It had an interesting program of research on agricultural products and chemicals of various kinds, including insecticides, herbicides, and applications that killed fungal diseases in cereals.

He talked it over with Margaret that night; but he had little doubt that he would accept. Margaret commented that, at its very least, the job was a launching pad into a career. Whatever he eventually chose to do, the experience would be invaluable.

He went to Wilmington, had a successful interview, and drew a small advance against his salary. When he returned Margaret took $300 from her savings and they bought their first car, a secondhand Pontiac that had been owned by her boss's father. They treated it as though it was a new Cadillac. She excused the extravagance by saying they needed transportation to Delaware. "Also," she reasoned, "we'll need a vehicle to look for somewhere to live."

They found their new home within the first week—a three-room apartment at 1307 Delaware Avenue. It had a separate kitchen, a bedroom, and a living-dining room, as well as a bathroom. It was compact and sparsely furnished, but they were immediately happy there. They made friends quickly, and Borlaug discovered he had a second cousin—an organic chemist with Du Pont, Gordon Vaala—living in the town. Young couples came into their circle through Borlaug's associations at the laboratory, and they settled easily into the new life. But it did not last long.

When the Japanese bombers shattered Pearl Harbor on the morning of December 7, 1941, Borlaug felt impelled to volunteer for the army. He was at once rejected and told that he had been classified under the wartime manpower regulations. His laboratory was to be turned over to special contract work in support of the armed services, and he would be kept in his position until further notice. He felt defeated at first, but there was no choice. Soon he and his colleagues were immersed in a wide spectrum of research and development, tackling all kinds of problems that would affect an army. They developed camouflage paint, protective aerosols, and chemicals to purify water. They also conducted research into the effects of fungal disease organisms on the deterioration of material in tropical war zones and studied the agents involved in corrosion and damage to precision instruments.

While he worked at Du Pont, the first DDT samples came into the laboratories. He saw the development of a crash program at the company's plant in Cleveland, Ohio, which was soon shipping 200 pounds of the substance each day. It went by bomber to the Pacific, where it was used to reduce the incidence of illness among fighting men subjected to malaria-carrying mosquitoes.

He enjoyed his work at the laboratory, and he and Margaret were happy. In their second year in Wilmington Margaret discovered she was pregnant. They waited excitedly for the arrival of the baby, trying out different names. As the time drew nearer, they finally decided on Norman if it was a boy, Norma Jean if it was a girl. The baby was born on September 27, 1943, a bright, blue-eyed, healthy daughter, and they quickly shortened the name to Jeanie. They were together now more than they had ever been before—and more than they would ever be again.

Four

The events that drew Borlaug to Mexico in 1944 had been set in motion before the war, when the Mexican government asked the United States for help in developing an agricultural program to bolster its foundering economy. The White House called on the Rockefeller Foundation, which agreed to study the situation, and in 1940 the foundation assigned a team of scientists to the project. Their objectives were to study the quality of the soil and to find ways of improving the crop yields of wheat, corn (a staple in the Mexican diet), and other grain crops. The three men chosen for the job were authorities in these fields: Richard Bradford of Cornell, an expert on soils; Paul C. Mangelsdorf of Harvard, a geneticist with special interest in corn; and Elvin C. Stakman of Minnesota, Borlaug's old professor, a specialist on fungal diseases in wheat and grain crops.

During the summer of 1941 the three professors spent weeks traveling over the impoverished Mexican countryside. On their return they submitted their findings to the Rockefeller Foundation in a massive report that became one of the classics of its kind—the first blueprint for the rural redevelopment of an entire nation.

Once it was decided to implement the proposal, it fell on

Stakman to pick the best man possible to head the project. The candidate would have to be a man of leadership, ability, and dedication—and, in a country where Americans were often regarded with suspicion, a man who inspired confidence. Stakman found such a man in Dr. J. George Harrar.

Young, physically strong, and intellectually tough, Harrar had been a protégé of Stakman in the early 1930s and one of his outstanding students. He had taken his doctorate at the University of Minnesota, but during his undergraduate days at Oberlin College he earned the nickname "Dutch," for Flying Dutchman, because of his exceptional speed in track. A fair-haired man, with broad face and serious gaze under heavy brows, he was a born leader and dedicated teacher. In addition, he had spent four years working on plant diseases and botany in Puerto Rico and had mastered the Spanish language. He seemed like the perfect man for the job.

The question was whether he would take it. Harrar was thirty-six; he had just been appointed head of the department of plant pathology at Washington State University, and had a promising university career ahead of him. He would have to weigh the comfort and security of his job at Washington State against the uncertainties of a grinding bootstraps operation in the backroads of Mexico.

Nevertheless, Stakman decided to approach him, and early in 1942 he traveled to Washington and presented his proposal. Stakman pulled no punches. He could offer neither security nor assurance of success—it was a high-risk venture that would exact a large measure of sacrifice. While Stakman talked, Harrar closed his heavy lids and reflected in his intense manner. Then he said: "I guess I might prove more useful as a scientist in Mexico."

It was settled. Harrar asked only that he be allowed to finish out the year at Pullman, and Stakman was happy to comply.

In February 1943 Stakman and Harrar set out on a three-month survey of Mexico to lay the basic groundwork for their research and development program. They used every form of transportation available, traveling by horseback and mule in places where wheels would get bogged down. They went from Mexico City to the far south, from the wet jungle area around Vera Cruz to Ciudad Juárez in the north near the border of the United States, exploring semiarid country, desert, and mountain highlands. From each of these regions they collected seeds—wheat, corn, beans, squash—which were to be the first test specimens of the program.

Back in Mexico City they hammered out a plan of operation, which they presented to the Mexican government. It called for the creation of a new organization, operating within the framework of the Mexican ministry of agriculture, that would be called the Oficina de Estudios Speciales, or Office of Special Studies. This was to be run by Harrar under the direction of the Rockefeller Foundation and in cooperation with the Mexican government. It was to be staffed jointly by Mexicans and Americans, and emphasis was to be placed on training the young graduates from Mexico's agriculture colleges—*agrónomos,* as they were called—who would be taught the latest American methods of crop production and disease control. It was hoped that the Mexican scientists would ultimately take over the program, at which time the Rockefeller forces would withdraw.

Once the plan was accepted, Stakman returned to Minnesota and left Harrar to the problem of searching out a task force. He would need specialists from the United States, but his first priority was to find someone reliable to handle the daily operations of the Office of Special Studies —"the Office," as it quickly became known.

His attention focused on a young agrónomo in the Mexican ministry of agriculture, Inginero José "Pepe" Rodríguez, who at twenty-three had recently graduated from a small national college at Chapingo, about twenty miles northwest of Mexico City, and was the first plant pathologist to join the agriculture department. He was bright, patient, and diplomatic; Harrar made him his right-hand man.

With the Office running smoothly in Rodríguez's care, Harrar returned to the United States in mid-1943 to seek out American scientists in his three major areas of concern: soil development, corn production, and disease control, particularly in wheat crops. The timing was not propitious. Wartime regulations made scientists scarce, and Harrar took his problem to two of the project's originators, Stakman and Mangelsdorf. Mangelsdorf suggested that he try Dr. Edwin J. Wellhausen, who had won recognition as a corn breeder in West Virginia. "If you can get this guy we'll be lucky," said Mangelsdorf. "He's got the right background. He'll set a plow, level a field, cut corn—darned near anything."

Harrar made inquiries and discovered that Wellhausen would be available. He went to interview him at his home and quickly decided he was the man for the job.

Meanwhile, others at the Rockefeller institution were looking for a soil specialist. There were a number of candidates, but the field was finally narrowed down to an agronomist from North Carolina, Dr. William Colwell, who had taken his doctorate at Cornell.

Harrar now had to fill the post of plant pathologist. Stakman highly recommended a young scientist with the Du Pont laboratories named Norman Borlaug but pointed out that Borlaug was restricted under the manpower regulations. Harrar was interested enough to pursue the possibility, however, and the two men cooked up a scheme to sound Borlaug out about his interest. Stakman was planning a symposium in Philadelphia that Borlaug would attend, and he suggested that Harrar come along.

During the informal discussion that followed the sym posium Norman Borlaug found himself next to a man whose face was vaguely familiar to him. They knew of each other but their careers had not brought them together until then. Harrar said: "You're Norman Borlaug. Let's have a drink."

The talk came easily. Harrar told Borlaug that he was heading the Mexican venture that Stakman and the others had originated; that he was looking for certain staff members—including a plant pathologist. The team was going to conquer hunger in Mexico, and the canvas of their operations was vast, the opportunities for worthwhile work enormous. "I know you're tied up with wartime controls," said Harrar, "and it's a pity. I feel we might do business."

He then asked Borlaug about his work at Du Pont and about his family. Borlaug told him about the laboratory routine and said that he and Margaret were enjoying the first months of their baby daughter, Jeanie.

But when he returned to the small apartment in Wilmington, he was in a pensive mood, thoughtful and quiet. That night he told Margaret about his meeting with Harrar and about the Mexican project. The restrictions binding him to Du Pont were beginning to chafe.

All that winter the Mexican project continued to nibble at the edge of his mind. One evening in the spring of 1944 Margaret answered the doorbell to find Dr. Harrar and another Rockefeller official, Dr. Frank Hanson, standing on the step. They had been in town on business and had called "out of the blue."

There was some idle talk and then the subject turned again to Mexico. Harrar became very serious. "Those people need help badly. And we need men with heart to give them that help." Then he said to Borlaug: "Would you come and join us if you were freed of your classification?"

Borlaug paused before answering. Fighting was fierce in the Pacific, the theater his work supported, and the war seemed to stretch ahead without end. The challenge of Mexico was attractive, but he had to consider the war and his family. He glanced at Margaret. Only a week before she had told him she was pregnant again.

"I'd like to consider any job that was really worthwhile," said Borlaug. "But I'm locked up pretty effectively."

"Look, Norman," Harrar persisted. "I'm simply asking if you *were* released from that damned classification—would you call me?"

Borlaug looked at his wife again. Margaret smiled, assuring him. "Sure," said Borlaug, "I'll do that." It didn't seem like a very real possibility at the time.

There was more than one reason for Harrar's visit that evening. He wanted to interview Borlaug again to be sure he was the best possible candidate. And he wanted to see him at home. He believed that a wife was important in keeping a team together, and he wanted to see whether Margaret seemed the sort of person who could stand up under the stress of the difficult living conditions and the long periods of separation that the job would demand. By the time he left Margaret's calm, self-assured manner and her obvious devotion to her husband had told him all he needed to know.

Unknown to Borlaug, Dr. Harrar and Dr. Hanson also paid a call on his employer, a chief executive of the Wilmington laboratories. Harrar told the executive that he understood Borlaug's present work was of military importance but that he was needed for a project of even higher humanitarian value. He mentioned that Borlaug was receiving a prewar salary of $2,800 a year, about half of what the company was paying recent graduates who were not under wartime controls. The executive remained adamant: the military deferment would stand.

Two months after Harrar's visit, however, the same company official called Borlaug into his office and told him he was to be released from manpower controls. The order would become effective the following month, but they did not want to lose Borlaug's services.

"I know the Rockefeller people want you for some scheme or other," he said. "But if you'll stay on with us, we'll double your salary."

Suddenly Borlaug was very upset—not at the knowledge that the war had been used to underpay him for the past three years, but at the inference that money would be more important to him than the kind of job he chose to perform.

That night he put through a long distance telephone call to consult with Stakman. In his recommendation to the Rockefeller Foundation Stakman had written of Borlaug: "He has great depth of courage and determination. He will not be defeated by difficulty and he burns with a missionary zeal." This was true, but only after Borlaug had decided on a course of action. Now he wanted his old professor's advice.

Stakman wanted Borlaug for the Mexican project, but he tried to be objective. "I have to tell you that it is a tough, demanding task you're being offered. It's a long, grinding project. But if I can encourage you, Norman, I'll say this: it would be a worthwhile thing to do, to put bread into those hungry bellies in Mexico."

After his talk with Stakman there was little doubt in Borlaug's mind that he should take the job. His only consideration now was his family. He and Margaret mulled it over. Du Pont offered security, a future; Mexico meant uncertainty, upheaval, separation, trials of adjustment. He would have to go ahead of her to make arrangements. When she needed him most, when the baby was on the way, he would be 2,000 miles away in a strange land. Even as they talked, Margaret had a feeling of inevitability. There was something ordained in the turn of events. His work had always been important; this task seemed to be supreme.

Yet he needed her assurance. She smiled and eased away his last hesitation. "You don't have to worry about me, Norman. You know full well I'm quite capable of taking care of myself. Now go and make that phone call."

He got through to Harrar in Mexico City, and Harrar was delighted—but not really surprised.

"Yes, I know you've been released," he said enigmatically. "We've been in close touch with the situation. How soon can you come?"

By the last week in September 1944 all the arrangements had been completed. Borlaug would drive down to the Texas border town of Laredo, where he would be met by Edwin Wellhausen, the corn specialist, who was already on the job at the Office of Special Studies. Together they would travel across the Rio Grande to Mexico City.

On a bright Monday morning the old Pontiac stood at the curb packed and waiting to carry him to Mexico. He ate an early breakfast and prepared to bid farewell to his wife and fourteen-month-old daughter, Jeanie. The decision to leave them behind had not been easy. But there was the language barrier, the strain of searching for accommodations in a strange country, and the uncertainty of medical services for the coming birth. Margaret was now six months pregnant and decided it would be better for her to stay among her friends in Wilmington until the baby was born. By then he would have a home for them and they could be together again.

He picked up little Jeanie and kissed her on both cheeks. Then, with a farewell kiss from Margaret and a final assurance that she would be all right, he was off.

He drove steadily southwest, thoughtful about the future, pensive at parting. But soon the changing landscape began to take his attention and he was absorbed in the adventure of the journey—through the Shenandoah Valley and the

Smokies, all aglow in their October splendor. It was the most beautiful landscape he had ever seen—better even than the autumn splendor of the Berkshires of Massachusetts. He allowed himself five days to get to the border—down through the mountains of West Virginia, through Nashville and Memphis, and across the wide, muddy waters of the Mississippi, so different from the swift, icy currents of the rivers near Saude. He made a leisurely stopover in Dallas, enjoying the open-handed Texan hospitality, then moved on to Austin, San Antonio, and finally Laredo. The little border town was burned with the autumn sun and bursting with the history and legend of the West.

The afternoon light hurt his eyes as it bounced off the whitewashed walls of the two-story buildings. He parked his car and walked along the short stretch of main street, past hitching posts and horses and tall, lounging Texans, to the Plaza Hotel, where he was to meet Wellhausen.

The hotel reception area was surprisingly cool, with potted palms and cane furniture. In the next room men stood at the bar in their tall hats, boots, and fancy spurs. As he made for the door, a tall, thickset man came toward him with arm outstretched. "You must be Norman Borlaug," he said. "I'm Ed Wellhausen." His manner was direct and warm. They walked into the darkened recess of the bar and chatted amicably over a cold mug of beer. They were both tough, determined men and liked each other from the outset.

Wellhausen's wife, Vivien, had come up from Mexico City with him, and the three of them had a hearty meal that evening, comparing notes on their journeys to the border. After dinner Mrs. Wellhausen excused herself early to get a good night's rest for the 800-mile trip back, and the two men slipped naturally into conversation about the Mexican venture.

Borlaug was eager for information about the Office of

Special Studies, and Wellhausen immediately launched into a discussion of the difficulties of dealing with the Mexican bureaucracy. Although Harrar was director of the project, the Office of Special Studies was largely staffed by civil servants of the Mexican ministry of agriculture, and cutting through the red tape was often time-consuming and frustrating.

"The bureaucratic setup there has been a pain in the neck," Wellhausen complained. "A tractor breaks down in the field and you need a paper order to get it fixed. Try to get it in a hurry and you haven't got a chance. The more you rant, the more your energy dissipates. It's like punching a featherbed."

The Office itself was housed in three hot, crowded rooms, allocated by the ministry of agriculture, in an old building on the outskirts of Mexico City in the San Jacinto area. There was no place that could be described as a modern experimental field station, and there was a dire shortage of trained Mexican agronomists. The embryonic national agricultural school at Chapingo, twenty miles northeast of the capital, had not yet produced a single M.S. or Ph.D. Surrounding the school were more than 800 acres of land that had been taken over by weeds. Harrar had been given 150 acres there to set up an experimental corn- and wheat-breeding station, but equipment was scarce and the only building on the land was an old shed with a tar-paper roof.

There were a few experienced men in the program, but they were working at a disadvantage. Edmundo Taboada, for example, one of the best men in Mexican agricultural science, had developed an experimental wheat nursery about an hour's drive from Mexico City. But he had to use hillside caves to store his precious seed and equipment, because he could not get storage buildings erected.

It was going to be an uphill struggle to get the project off the ground. And yet, explained Wellhausen, it was not for lack of good will at the top. Dr. Marte R. Gómez, the Mexican minister of agriculture, and his sub-secretary, González Gallardo, were eager to cooperate. But the machinery of bureaucracy ground slowly. Borlaug went to sleep that night with Wellhausen's words troubling him. He had not counted on the irritating details of operational problems.

They drove out of Laredo soon after sunrise, Borlaug following the Wellhausens in his Pontiac. As he crossed over the border into Mexico, he had no inkling that America would never be his real home again, that he was embarking on a venture that would touch a billion lives across the globe. He was aware for the moment only of the dust from the wheels of the car ahead, the heat of early morning rising from the dry land, the flat plains of arid desert stretching away from either side of the road.

The long drive gave Borlaug a chance to soak up the color and variation of the Mexican landscape: the foothills of the Sierra Madre, dotted with fertile forests and laced with trickling streams and waterfalls, wide stretches of scrub desert, and an occasional Indian village, like an oasis of habitation. South of Ciudad Victoria the villages were more numerous, the pastel walls splashed with the vivid color of ornamental vines and clusters of zinnias and poinsettias. There were mango groves, banana plants, orange trees. Driving steadily southward, they moved into the semitropical zone of the Tropic of Cancer and faced the long climb over the Sierra Madre Oriental. From the heights of the Madre Borlaug could look down on the vast panorama of the central mesa. Then began the slow descent to the urban and industrial areas, and finally the route through the crowded, noisy streets of Mexico City to the suburb of San Jacinto and the Office of Special Services.

The project director, George Harrar, was there to welcome him with a strong handshake and broad grin. He was introduced to William Colwell, the soil scientist, with whom he would be working closely on soil fertility and improvement. He had just missed Harrar's young Mexican assistant, José Rodríguez, who had gone off to Minnesota to study for a year. "He'll come back another Stakman product," said Harrar. "And there'll soon be other graduates joining the program. I'm picking the cream of the crop, but we can't hope for too much too soon. Our first problem is to inject a little modernity into this out-of-date agricultural system."

He showed Borlaug around the cramped offices and gave him a brief rundown of the immediate problems. It confirmed everything Wellhausen had told him. Facilities and equipment ranged from meager to nonexistent. The Rockefeller Foundation could not help with trucks or jeeps because the U.S. government had put wartime priority on all such vehicles. Gasoline and spare tires were nearly impossible to obtain. Building materials, farming machinery, and irrigation equipment were in equally short supply.

The science library Harrar had requested was yet to come, and it was urgently needed in the battle with the many little-known insects they encountered. Borlaug's skill as a plant pathologist would be tested to the utmost in the daily struggle against a wide variety of tenacious fungal diseases. There was a mountain of work and too few people to do it.

By the time Borlaug reached his hotel room in Mexico City—the Hotel Geneva, which was to be his home for the coming weeks—his head was reeling. He was bone-tired, and the task before him seemed formidable. What kind of future was there in this strange country? How could they hope for a major revolution in agricultural production when they could barely implement the most fundamental agrarian reforms?

Five

Revolution was not new to Mexico, nor were attempts at rural reform. Following the bloody struggle which opened in 1910, some 135 million acres were stripped from the great landowners—the Roman Catholic Church and the formerly powerful hacienda families—and turned over to the *campesinos,* the small farmers. This land redistribution program was known as the *ejido* system—land for the comman man, land for those who worked the land. For 2 million peon families it brought hope of a better and more plentiful life.

But the ejido system created problems of its own. Land was parceled out according to population density, locale, and the family's role in the revolution. The lots ranged from four or five acres, often on slopes too steep to till properly, to blocks of hundreds of acres of fertile soil. For more than a million peasant families the hope of a plentiful life gave way to harsh reality as the worn-out and unproductive earth held them in savage thraldom. Moreover, many of the new inheritors were not prepared for their new role. They lacked skill and experience, even hand tools needed to break open their new territory. Many had never grown more than a ragged patch of corn. To help them, the government set up ejido banks to make loans for seed, animal feed, and equipment, but all too often the money was squandered away for lack of the most basic knowledge of animal husbandry and crop production and rotation.

The revolution of 1910 had created a new class of land-owning workers, but after thirty years of trial the reforms failed to end poverty, ignorance, and hunger. It was in this atmosphere of apathy and malnutrition that Borlaug joined the small force that was to create the first bloodless revolution in Mexico's history.

Norman Borlaug spent his first weeks in Mexico working in the experimental plots at Chapingo. He and Harrar concentrated on wheat, Wellhausen on corn, and Colwell on soils. The basic approach was simple. They selected promising plants, on the basis of yield and apparent resistance to disease, and planted them again for cross-pollination. Tests were made to measure the response of the different specimens to various soil compositions and fertilizer applications.

These experiments were limited, however, to what could be done in the given climate and locale of the two experimental fields. In order to broaden their base of operation, Borlaug and Wellhausen decided to journey into the great central highland country known as the Bajío. They hoped to persuade local farmers there to allow them to plant test strips of corn and wheat on their land. This not only would give the project added land facility but would help open farmers' eyes to the fact that a scientific approach to farming would eventually bring profit. It would also test the response of different varieties of seed to various altitudes—and in Mexico altitude is an important factor.

The Sierra Madre mountain chains create a bewildering assortment of climates and environments. The two great mountain ranges run like jagged backbones along both sides of Mexico, east and west, each more than 1,000 miles long, each parallel with the sea. They lift the entire hinterland into a central mesa, from 3,000 to 8,000 feet in altitude, with wide variations in soil, topography, and temperature.

It was late in October when Borlaug and Wellhausen went rattling up into the highland country northwest of Mexico City in the old Rockefeller pickup truck. Here in the Bajío, in the states of Querétaro, Guanajuato, Michoacán, and Jalisco, Mexico's traditional breadbasket, Borlaug first came face to face with the shocking pauperism that antiquated agriculture inflicts on an expanding population. He had been warned what to expect. Stakman had told him of his own shock on visiting the Mexican countryside. The Office scientists at San Jacinto had also told him of the poverty. And Wellhausen particularly had outlined in detail the problems he would face.

Wellhausen, in charge of the corn-breeding program, faced a monumental task himself. Corn covered 10 million acres of Mexico's farmland—ten times the area devoted to wheat—yet it did not feed the country. Mexico was importing tens of millions of bushels of corn and wheat each year, and wheat imports alone were costing 100 million pesos annually—then about $30 million. This reduced the currency available for essentials like machinery, power-generating equipment, chemicals, and dozens of other commodities vital to the national welfare.

Borlaug knew all this. Yet out on the Bajío the reality of the situation was an appalling shock. The land was baked dry; bare rock protruded where storms had washed away the topsoil. Thousands of years of planting the same crop season after season had sucked the guts from the land. The life-giving minerals of the soil—nitrogen, phosphorus, potash, copper, cobalt, zinc, molybdenum—had been mined from it by countless plantings. The vital elements had been taken without the farmers' knowing they existed, and there was no thought of replenishment. Enrichment of the soil was unfamiliar. Fertilizer was largely unknown, and even where it was understood it was too expensive.

There were some good farms, where progressive farmers used animals and machinery to advantage, but they were few. Overall, the depleted lands imposed a poverty that was heartrending. It was the plight of the people that moved Borlaug most. Hopelessness pervaded their lives. He could see it in their actions and their faces.

One evening, as Wellhausen cooked their dried meat and boiled the coffee, Borlaug sat leaning against a fender of the pickup and penciled a note to Margaret:

"These places I've seen have clubbed my mind—they are so poor and depressing. The earth is so lacking in life force; the plants just cling to existence. They don't really grow; they just fight to stay alive. Nourishment levels are so low that wheat plants produce only a few grains, and even the weeds and diseases lack the food to be aggressive. No wonder the people are the way they are! Can you imagine a poor Mexican guy struggling to feed his family? I don't know what we can do to help these people, but we've got to do something."

Borlaug was not in the best condition himself by that time. For the first two weeks on the road he and Wellhausen had stayed in small hotels and boarding houses in the small Bajío towns, sleeping in insect-ridden rooms. Lack of hygiene, and poor water and food, had made them ill—as well as highly uncomfortable. Thereafter they lived outdoors, sleeping in barns or sheds or under their truck. They boiled their water religiously, but Borlaug was still sick. Wellhausen became sick too, but Borlaug was hit harder. Enteritis seized him in its agonizing grip, and he weakened with dysentery and nausea. Some nights he writhed in his bedroll, swearing he was mad ever to have left the laboratory job in Wilmington. He did not mention his illness in his letters to Margaret.

They spent several weeks in the Bajío and the surrounding mountainous areas, traveling hundreds of miles over roads unworthy of the name. There were trails of powdery dust that turned to slime in sudden downpours and iron-hard washboard roads that rattled their teeth and shook their bodies until they ached. But Borlaug forgot his own weakness and discomfort as he looked at the scenes of human misery around him.

In some of the mountainous areas, the people worked patches of corn, more often than not on steep slopes, their bodies angled away from the incline. Women sometimes clung to the bottoms of cornstalks to keep from slipping. There were many Indians, quiet, introspective people, and the heaviest weight of poverty seemed to bear down on them. They had no collective voice that Borlaug could detect. They lived in isolated pockets and spoke in many different dialects. Spanish was unintelligible to them, a foreign tongue that made them suspicious and wary.

The Indian children aroused intense pity. They were timid, awed by the truck and the strange scientists. At their approach they would scurry away like insects, peering at them from a safe distance with dark, sunken eyes. Their little bellies were distended, their legs thin and knotted from lack of protein. Adults and children—all reflected the same condition as the played-out soil.

There were tall, high-cheekboned Tarascans, an industrious, dignified people. The Tarahumaras, a stockier, sturdier people, were close to primitive life: shy, reserved, unwilling to communicate with strangers. There were Otomi villagers and Huastecs and many settlements of mixed Indian-Spanish blood. Each had its distinctive culture, architecture, and language, but all had something in common. They did not have enough food and they lacked education.

In the 1940s rural education in Mexico was a national disgrace. Illiteracy was rampant. An official survey revealed that four out of five of the rural populace could not read or write. Only 3 percent of the rural schools offered more than six years of education. The remaining 97 percent offered four years or less, and many schools in the Indian villages were shut for long periods because there were no teachers.

What schools there were were designed to train peasant children in farming skills and to try to improve life and living standards, but they failed miserably. The courses were devised in departmental corridors of government establishments in Mexico City. They were often impractical and rarely backed with adequate funds. As late as 1944 a noted Mexican educator and reformer, Dr. Jaime Torres Bodet, wrote of the country's rural schools: "Their condition is pitiful. The students have no clothing. The bookshelves do not contain any books. And why speak of workshops and laboratories when, in many instances, we have not even been able to provide farm implements, or even farm animals, with which the students might work." Researchers at the Rockefeller Foundation had found these conditions to be an essential part of the Mexican malady. They noted that even if the schools had been well equipped and staffed, and even if the children had clothing and books on the shelves, the effects would have been minimal because the teaching methods were obsolete.

As he went through the countryside, Borlaug saw that antiquated theories of farming were perpetuated from one generation to another. Even on the bigger, richer farms the foremen—*mayordomos,* they were called—were invariably hostile to new ideas, particularly when espoused by gringos. Many farmers still held to myth and superstition.

Borlaug's anger grew as the miles rolled away. Vignettes remained fixed in his mind for life: a peon in a dirty white poncho shoveling wet earth onto his leaking dirt roof; a woman in a shapeless black dress holding a baby to her breast with her left hand while swinging a sickle with her right hand in the midday sun. He fumed at the waste of life, the inhumanity. The indifference of remote bureaucrats, overcautious scientists, and academics to these conditions produced in Borlaug a lifelong enmity toward officialdom. These people needed help, guidance, counsel. They had to be shown how to manage the land, control the pests, and guard against disease. New crops with higher yields were essential; enrichment of the aged soil was vital. But officialdom offered neither help nor hope.

The nation was paying more than 100 million pesos each year for wheat grown in foreign fields. Suppose that grain could be grown in Mexico's soil and the pesos put into the pockets of Mexican farmers? What a difference it would make to their lives and to the whole economy of the country.

It was mid-November by the time they returned to Mexico City. As Borlaug walked into the cool foyer of the Hotel Geneva, he suddenly realized how happy he was to be getting back to hot water, clean sheets, and good food. He was haggard and thin from his bouts of sickness, fatigued from weeks of travel and rough living. He was glad Margaret was not there to see him. She would have been concerned.

When he picked up his key from the desk, the clerk handed him a slip of paper. The message had been waiting there for two or three days, and the first words wiped all thoughts of fatigue, sickness, or Mexico's troubles from his mind. There had been a telephone call from America.

Margaret had given birth to a son; she was well but there was some trouble with the baby. A letter was on the way.

His first thought was to call Margaret. It took an hour to get through to the hospital in Wilmington, but Margaret was sleeping in a ward without a telephone and they would not wake her. They would give no information over the telephone. He would have to call back the next day. He lay down on the bed and spent a fitful night, worrying about what to do next, wondering what was wrong.

The letter from Margaret arrived the next morning. She had written under great stress, he could see that from the stilted, broken sentences. Two strange words caught his eye, "spina bifida"; and though she wrote of a spinal deformity, and of an operation that might be very dangerous, he did not understand what it all meant. He telephoned Dr. George C. Payne, Rockefeller's resident medical man in Mexico City. What was spina bifida? He told him why he was asking and Payne's voice went immediately quiet and grave.

It was a spinal deformity, probably a genetic aberration. Usually it meant that the spinal column, the nerve cord, was without protection because of a gap in the vertebrae. In some cases it was severe and the nerve cord was exposed. There was danger from sudden movement and the risk of infection. It almost certainly would mean paralysis of the lower limbs and functions, and might threaten other faculties. "The severity of the condition varies widely, Norman," Payne told him. "It's hard to say without knowing. I'm terribly sorry—but I have to say that at its best, it is bad. At its worst, it's tragic."

He saw all too clearly what Margaret faced. He had to get to her. He went to the Office and tried to get his old Pontiac, but the car was now under bond to the Rockefeller institution and could not be taken out of the country without Harrar's authorization; the director was somewhere in

New York and could not be reached. Borlaug thought of taking a plane, but he had no money to fly, and again Harrar was the only man with access to Rockefeller funds. The bureaucratic rigidity was maddening. He sat in one of the Office rooms and called the hospital. This time he got through to Margaret. She was controlled and matter of fact about the situation. They were considering an operation that could be corrective, she said, and the baby had the best doctors available.

Borlaug was deeply moved by her courage. He apologized and explained that he couldn't get out of Mexico City until Harrar got back. "That's O.K., Norman," she said. "Don't worry too much, it won't help. I can cope till you get here. But come as soon as you can."

It was a week before Harrar came back. Borlaug was immediately advanced the fare, and he took the first plane he could get. Margaret was out of the hospital by then and at home with Jeanie. He called her from the airport, and when the taxi brought him to the door she was almost distraught with relief. They fell into each other's arms and in the following hours bolstered each other with words of faith, consolation, and comfort.

"I made the decision about the operation, Norman," she told him. "There was no more time. They wanted to cut the poor little thing up, but the way they explained it made it sound very desperate, very little chance, and I felt it was better to let nature take its course. The operation was too much like a way to end it all. It was just a gamble with his life."

The following days and nights were full of worry and anxiety. There were conferences with the doctors, counseling with the pediatrician and other specialists. Nothing seemed conclusive. It was as though everyone was waiting for something to happen.

It was the worst for Margaret. She had named the baby

Scotty, and she had great dreams for him. Now she had to face the probability that she would never hold him, never touch him, never know him. Her baby could be handled only by skilled, trained people. It could be weeks, months, years. What was most to be feared—and expected—was hydrocephalus, water on the brain. That would mean paralysis, loss of faculties; the child would be a vegetable, institutionalized for the duration of its life. It was a savage fact to face.

Under the stress of their grief, Borlaug took to walking from room to room, tapping his fingers, pursing his lips in long silences, wandering down to the corner store for things they did not need. He came back from one of these walks and looked hard at Margaret with his head down, peering from under his brows as he did when he faced a difficult decision. He had run into a senior executive of the Du Pont Company. The man had heard about their trouble and offered him his old job back. The door stood open.

It forced another dilemma on them. "If you want us to stay together in Wilmington and be close to Scotty, that's what we'll do," said Borlaug. "I can take the Du Pont job. I don't have to go back to Mexico."

But Margaret had no illusions. She knew her husband—probably better than he knew himself. Mexico had struck an incredibly strong note of challenge in him. Not only did he want subconsciously to go back and work there; he needed to go back. He knew he could do something constructive there.

She held out her hand to him. "My husband has a future," she said. "My baby has none."

Borlaug returned to Mexico in the winter of 1945. He hated to leave Margaret, but he knew she was right. He could do nothing to help the child by staying.

Back in the Office at San Jacinto Harrar was there to

welcome him, a sympathetic hand on his shoulder. But the director turned quickly to the business at hand, perhaps sensing that hard work was the best antidote to his colleague's grief. "We've a big year ahead, Norm. The war's coming to an end; we'll get more people, more money, more transportation and equipment. We've got some extra land ready to work at Chapingo—150 acres—and plans for a new building. Stakman and Mangelsdorf are looking for new people in the States, and we're getting more agrónomos to work with us in the fields."

They held meetings—Harrar, Wellhausen, Borlaug, and Colwell—to discuss their plans. But Borlaug felt stifled in the musty, high-ceilinged rooms of the Office. It was crowded with desks, files, and cartons of paper, the floor littered with samples of plants and packets of seeds. He picked up some of the seed envelopes and saw from the data Harrar had collected that a wheat-breeding program was already under way.

He was itching to get out and do something active, and after one of the staff meetings he said to Wellhausen: "I need air and sun and land. Let's go talk to some of those farmers, Ed."

They set out for the Bajío again, back over the deep-rutted roads they had traveled in October. The winter wind of the plateau was chilly and crisp, and Borlaug felt it burning his face, easing away the stress of the tragedy in Wilmington. It would not blow away the ache he felt for Margaret and his baby son; but work, challenge, the open air—this was therapy.

Again they slept out at night, avoiding hotels, carefully boiling all their water, and eating from a stock of dried beef and biscuits they brought from the city. It was good nutrition, but monotonous and uninspiring. After the first week on the road they talked of food and re-created dishes they had eaten in the past. Borlaug had a recurring dream of eggs and bacon sizzling in the pan and thought again of his grandmother's fresh-baked bread.

They revisited the farms where they had planted experimental plots of corn and wheat, pushing northwest from Puebla to Pachuca, Querétaro, Celaya, León, and on to the broad highlands beyond, as far north as Saltillo. They worked ceaselessly, searching for clues to the response of each different seed type to diverse environmental pressures.

Borlaug inspected the wheat for the brown pustules that threatened devastation. If the weather turned moist and warm, these small pustules would break and release infectious spores of rust disease to float on the wind. One single rust pustule could release a quarter of a million spores; an acre of infected crops might harbor millions upon millions of such spores. And every spore was a threat.

The native wheats of Mexico—those they called criollos—had long ago lost all thir defenses against rust infections. Unless a cereal has developed the structural chemistry to bar entry of the spores, they penetrate the stems or leaves and suck out the plant's moisture and rob it of nutrients. In so doing, they also clog the plant's feeding channels and break down its strengthening fibers. Rust has many varieties—the main ones being stem rust, yellow rust, and leaf rust. But each variety breaks down into different parasitic strains, or races; all look alike under a microscope, but they are lethally different in the manner of their attack on the growing plant. More than 300 races are known in wheat stem rust alone, and any race can quickly mutate to destroy crops that hitherto have been immune to it. Thus there could never be any certainty that wheat specially bred to withstand the most prevalent strains of rust one year would be immune the next. The struggle to keep ahead of the disease was unceasing.

In Mexico there had never even been an attempt to control rust infections. When conditions favored the rapid spread of stem rust—and its associated diseases and blight—miles of crops would turn brown and die. Then growers faced total devastation, and to make money farmers had turned away from wheat to corn and other crops. In some years stem rust had repeatedly swept through the major grain-growing areas of Mexico. In three successive years, from 1939 to 1942, it had slashed the national grain harvest in half.

During his second journey to the Bajío Borlaug tried to talk with Mexican farmers about rust, and though he caught the sense of horror that they associated with the disease, it was impossible to communicate properly. Language was a barrier, but it went deeper than that. There was the more difficult, impenetrable barrier of suspicion. Borlaug and Wellhausen would drive up in their battered pickup, and a farmer would look up from his rows of corn and wait for them to speak. But he would not really listen to what they were saying. He would say nothing, do nothing.

Wellhausen knew the symptoms. It was the passive resistance, built up over decades, to the government man, the official who came in his white shirt, clean boots, and motor car and who took but never gave. "These people are steeped in suspicion," said Wellhausen. "And with good reason. But goddamn it, we're going to have to be tenacious and get them to trust us enough to listen."

Sometimes a farmer would let them sleep in a vacant room or shed, and they would be up and away at daylight. It was wise not to take advantage of the concession; it was a bridge to future contact.

At the end of February they started back, and by early March they had reached Chapingo. Borlaug did not continue on to Mexico City with Wellhausen but stayed to look over the experimental wheat crops in the area.

Harrar had laid the groundwork for the wheat project the year before Borlaug arrived in Mexico. He had collected seed from various sources, always with an eye to built-in resistance to rust. He had planted several hundred varieties, mostly at Chapingo, but a few out in Puebla. When they were young stands, he had deliberately infected them with the pathogens, or spores, of stem rust. Later he had harvested seed from the plants that showed the highest degree of resistance. These he would plant the following season. It was a process of elimination.

In his first months in Mexico Borlaug had worked closely with Harrar in the two fields at Chapingo. Now he was curious to see how this wheat, planted in October and November, was faring. He slept in the shed and was at work among the rows of test plantings at dawn. There was just enough evidence of stem rust for him to distinguish between plants that had genetic resistance and those that were vulnerable. But it was not a see-at-a-glance job.

They had planted a hundred varieties of wheat in 1,800 rows, each about fifteen feet long—a walking distance, plant to plant, of nearly five miles. To inspect the plants, Borlaug doubled over into a crouching position and peered under the curling leaves, fingering the stems, searching for the brown spots of infection. As he went from row to row, he jotted down details in a notebook: which types showed highest resistance, which were already heading with spikes of grain, which were late. It was hard work, but gratifying—the sun biting his neck, the earth fine and friable and staining his hands, the smell of soil mingled with the smell of dew and young wheat.

That was how Harrar found him one afternoon when he went to check on Borlaug's progress. "Norman!" He spoke a little abruptly. Borlaug raised his head. There was the project director purposefully striding down the rows in his leather knee boots, cavalry pants, and khaki shirt. Harrar came quickly to the point. "I've got too much to handle. I want you to take over the wheat program. Organize it and run it any way you want. I'll give you all the help I can—when I can."

A few more words, a brief discussion on the condition of the wheat, and Harrar went back to Mexico City and his increasing worries at the Office. It did not occur to either of them to speak of a raise in salary for the new responsibility.

It was a real problem to unload on Borlaug. There was no machinery, equipment, or transportation, and wheat was relatively new to him. He had not worked much with it on the family farm, and his doctoral study concerned flax and related rust diseases. But not wheat. Now he had two fields containing perhaps 30,000 individual strains. He had to know them, each and every one, intimately.

Borlaug commuted to Chapingo for the rest of March and into the first week of April, going back to his hotel room in Mexico City for clothes, a change of diet, and a hot bath when he could hitch a ride. Otherwise he spent all the daylight hours in the fields.

The idea of tall wheat being symbolic of productivity had been accepted by the world's farmers from the beginning of history, and the varieties at Chapingo gave what then was considered a good grain yield for Mexico, about eleven bushels an acre. But tall wheat also had failings: it could not be fertilized into greater productivity without the risk of growing spindly and weak and falling to the ground, a condition known as lodging. This meant that extra feeding would often result in less grain, not more.

At Chapingo Borlaug developed an intimacy—almost a sixth sense—in dealing with his wheat. He found that in its manner, its feel, its movement, a plant could talk to him, tell him of its growth and health. It amused his Mexican colleagues, but later some of them came to take the idea seriously.

As his knowledge grew, Borlaug became increasingly

certain that no single wheat was perfect for Mexico. The problem was twofold: the many different climates, elevations, and environments, and the endless adaptability of rust to new strains. One day several staff members of the Chapingo college visited him at the experimental grounds, and he talked to them about this concept.

"There are thousands of wheat plants here," he said. "Each head will grow a couple of dozen grains of seed—and there will not be one seed in half a million that will be perfect for what we need in Mexico. Perfection is a butterfly the academics chase and never catch. If we go on looking for the ideal wheat for Mexico, your countrymen will go on being hungry for a long time. We will have to do the best we can with what we have."

Borlaug was aware of the time factor involved. It usually took between ten and twelve years to breed a new, improved wheat and to put it into production in a specific locality. This was because of genetic law, which dictated that a cross between two wheats could be judged not by the characteristics of the parents but by what they produced. And what they produced would not become truly apparent until two further generations were obtained from subsequent plantings of the seed. This meant that the results of any given crossing could not be determined for three growing seasons, and each subsequent improvement would take another three years. The final step would be multiplying the seed necessary to supply outlets.

With the threat of rust epidemics always on the horizon, the time element was a nagging concern. It seemed almost impossible to grow enough wheat for Mexico's bread.

Shortly after his talk with the Chapingo group, one member of the college staff returned with his brother, a farmer who wanted to ask some questions about wheat for his land. Borlaug shook hands with the tall, dignified Mexican and felt a calloused palm that told of hard work in the fields. Roberto Maurer was in his late twenties, handsome, with a thick, curving mustache. He spoke faltering English in a modulated, pleasant voice, carefully choosing his words, his dark eyes serious, his manner courteous. Borlaug liked him at once and listened to Roberto's tale intently.

Maurer was a scion of one of the old hacienda families of Spanish descent whose land had been stripped from them under the agrarian reforms set in motion by the revolution of 1910. They had pleaded for compensation for many years, and in 1936, Roberto had been awarded a tract of 250 acres of virgin cactus and mesquite desert land. It lay in the far northwest of the country, near the town of Obregón in the state of Sonora, between the Sierra Madre and the Gulf of California—the stretch of water between the Mexican mainland and Baja California. Roberto had cleared the land of cactus and mesquite by hand, acre by acre. His young wife, Teresa, a trim Spanish beauty, had joined him once he had built a temporary shelter, and they had followed the example of other farmers, putting in wheat as irrigation became available. In 1939, after three years of work, they had ten acres of wheat, which promised a good harvest. The crop was a conglomeration of several varieties, as was common in Mexico. There were native criollos, an American wheat called Barrigon, and other American strains from Texas and Arizona.

The wheat heads were filling out when a rust epidemic struck, and all across Sonora tens of thousands of acres of wheat were laid waste. Within days Roberto's ripening wheat had collapsed and the field became a tangled mess of matted straw, gray-brown and rotting in the sun. Teresa cried bitterly.

There were only a few clumps of wheat that did not fall into ruin. Roberto recognized them as the Barrigon and collected enough seed from them to plant the following fall. But the rust had struck again the next year and again in

1941. During the three years of rust epidemics only the Barrigon had shown resistance, and then not all the time. Roberto had continued to grow Barrigon, but this was a poor-quality wheat with a moderate yield. What should he do? Go back to higher-yielding varieties and risk the threat of rust, or grow poor-quality grain? Did science have an answer?

Borlaug in fact had no answer for Roberto Maurer; the best he could offer was hope. "We're working on this problem," he said. "And maybe we'll come up with something that will help."

Then he recalled a fragment of conversation he had picked up somewhere. "Tell me, Roberto. Isn't there some sort of government experimental station up there near Obregón?"

Maurer shrugged his shoulders. A model experimental station had been set up in the mid 1930s, he said, by the son of a former president, Don Rodolfo Elías Calles, when he was governor of the state of Sonora. But because of lack of money and trained personnel it had been handed over to the federal ministry. It was near his own land, in the Yaqui Valley, named for the Yaqui Indians, who had hunted there for centuries. But the station had long since been essentially abandoned; it might as well not exist.

They discussed soils and fertilizers for a while, then Maurer shook hands and left. Borlaug pondered over what the Mexican farmer had told him. He had heard of the wheat lands of Sonora and the neighboring state of Sinaloa, and suddenly he had a great desire to see them himself.

He hitched a ride to San Jacinto and called on Wellhausen. He related his talk with Roberto Maurer and how he felt compelled to look into the situation up in the northwest.

Wellhausen was not encouraging. "Listen, Norman," he said. "We know all about Sonora from the ministry. Good farms up there, doing well. They get double the yields recorded elsewhere. They can afford machinery, equipment. They're the last people in Mexico that need our help right now."

Borlaug was not convinced; he went to Harrar. "This darned place is chewing at my mind like a hungry rat. I think I'll take a trip up there while I'm waiting for the Chapingo wheats to ripen. I ought to know what the hell they're growing up there. What their problems are."

The Sonora town of Ciudad Obregón, at the southern entrance to the Yaqui Valley, could be reached from Mexico City in three ways. The overland route was totally impractical. It meant 1,200 miles of rugged cross-country travel, and there was not that much time to spare before he had to be back at Chapingo. Rail was also slow. The third possibility was the most attractive. A lumbering, privately owned six-passenger plane ran a shuttle service twice a week from Mexico City to Ciudad Obregón and could get him to his destination in two days. He took the first flight he could, touched down for a night's stopover at the coastal town of Mazatlán, and finally landed on a narrow dirt strip cut between wheat fields a mile from Ciudad Obregón. A cranking, wheezing green bus took the flight crew and passengers the rest of the way into town.

It was a bright, clear day in April when he stepped down from the bus onto the main street of the town. The faded single-story buildings and wooden sidewalks formed a frontier setting. In the unpaved street desert winds stirred whorls of dust that coated the faces of the buildings, the windows, and the people. Farther back from the main street were rows of adobe dwellings and some brick buildings; dark doorways looked in on one-room huts with earthen floors.

Borlaug picked up his bedroll and a small pack of canned

food and walked on through the town to the road running north. There he hitched a ride with a Mexican truck driver in the direction of the old experimental research center. He spoke Spanish well enough only to make the driver understand where he wanted to go, but the man talked on, grinning, the words spilling out, beyond comprehension. But Borlaug grinned back and nodded. "Sí, sí," he said every now and then.

Soon the driver tired of the monologue, and Borlaug contemplated the landscape in silence. Ahead the country opened out as far as the eye could see, gently sloping, ideal for wheat and for irrigation. Away to the left the distant Sierra Madre rose out of a hazy landscape, lifting broken edges of its fissured blue-metal face to the sky. High behind those peaks was a great dam, which had been built in the 1920s and 1930s to store the rain when heavy summer storms roared in from the Pacific. Water from this reservoir flowed down canals and through ditches that crisscrossed the valley and made the crops—particularly wheat—responsive. In the first years of irrigation, Borlaug had learned, some of the farms had achieved yields of forty bushels of grain an acre, as against the national average of eleven bushels. These yields had dropped, because of consistent cropping and poor management, but it was still good land and the soil had not been denuded.

From the truck Borlaug could see the water gushing along a canal at the side of the road, spilling off into ditches in furrowed fields to feed the roots of valuable commercial crops. It was a wonderland after the hard, flinty soils of the central mesa.

The truck jolted to a halt by the side of the irrigation canal. The Mexican driver smiled and pointed to where two stone pillars supported a rusted wire fence. Borlaug thanked him and hopped to the ground, turning away to keep the dust of the departing truck from his eyes.

When he saw what lay behind the fence his heart sank. A once beautiful research station was a shambles of neglect. The lawn in front of the buildings was overgrown with weeds. Fruit trees were diseased and unpruned; shrubs were unkempt and straggling. The experimental fields were a tangle of uneven wheat plants and weeds, with saplings growing up from the bottoms of water canals. The fences were fallen or ripped away.

He walked to where the deteriorated buildings stood in the late afternoon sun. Doors hung askew; windows were shattered, the wire screens ripped from the casements. There were great gaps in the red tiling of the roofs. Inside broken and rusting pieces of machinery shared the cluttered spaces with rotting sacks and decayed vegetable matter. It was an affront to his sensibilities.

This neglected shambles had once been a man's dream, the brainchild of Don Rodolfo Calles, who had planned it—almost forced it into existence—to help farmers in the area produce more food. It had been stocked with cattle, hogs, poultry. There were pleasant homes for the staff and a main building with a communal dining room and kitchen. But the Sonora authorities had found it too expensive, and the federal ministry in Mexico City, to which it was turned over, had let it slip into disrepair. Officialdom had little interest or know-how to help the distant farmers of the Yaqui Valley.

The two agrónomos who were in charge of the station were away, and an old Mexican, who was paid a pittance, looked after the place. Borlaug found him in a corner of the main building sheltered from the afternoon sun, his face in the shadow of an old sombrero. He tried a few words of Spanish, but the old caretaker could not understand him. "No le entiendo, señor."

Borlaug went into the communal kitchen, but it was too filthy to use. He went back to one of the smaller houses, arranged a few stones, and lit a fire to heat a can of beans. He unrolled his bedpack on the stone floor and retired early. During the night rats ran over his face, so he pulled a coat over his head and slept restlessly until dawn.

At light he was up, ready for a tour of the valley. He visited the local farms, collecting heads of wheat, which he carefully labeled and stashed in his bag. He looked for evidence of rust and asked the farmers questions. He heard again the story of the three years of rust epidemics. But now the valley looked prosperous, and he saw machinery and equipment that the farmers in the Bajío could never afford.

He was back at the station in the late afternoon, thirsty and hungry, covered with dust and sweat. He found the old caretaker, who showed him where to draw water from a pump well. He washed in the cool water and changed his clothes. When he looked up he saw the Mexican watching him with curiosity.

"I'm going to come back here, amigo," he said, making signs to illustrate the sowing of seed. "Six months. Plant some seed. Yo comprender, sí?" he said in his fractured Spanish.

The caretaker showed his stained teeth. "No le entiendo, señor."

Borlaug spent three days touring the valley. On the morning of the third day he found Roberto Maurer working his wheat fields and clearing more land. The astonished Mexican could not believe his eyes. He insisted on inviting Borlaug to his house for a meal. Smiling and delighted, he ushered him into their small brick home to meet his dark-haired, attractive young wife, who greeted him warmly. Over lunch the three of them began a lifelong friendship.

After their meal they discussed Borlaug's plans to return in the fall. Then Borlaug made the slow, two-day journey back to Mexico City. Sitting in the old airplane, he looked down again at the wheat fields on the Yaqui Valley farms and fingered the wheat heads he had collected. He turned over in his mind the concept that had taken shape from the visit. He had much to do, much to plan. But he was not yet ready to discuss his ideas with anyone.

Six

When Borlaug returned from the Yaqui Valley, once again he found a letter from Margaret waiting at the hotel. The last hope for their son, Scotty, had faded. Margaret had been visiting the hospital regularly, twice a week, only to peer at him from behind the glass enclosure, and each visit had been harder to bear. Then, at the end of February, the physician had called her to his office, and she knew the worst had come.

Hydrocephalus had developed, as had been feared. Life might last a month or two, a year at most, but there was nothing more to hope for. The doctor had patted her on the hand and urged her to accept the situation, however difficult. Visits to the hospital would only become more harrowing and psychologically damaging. "I really think," he had said kindly, "that it is time you went to Mexico to take care of your husband."

Her only thought was: "Please, God, be good. Take him soon."

Circumstances were choking her resolution to stay with her baby until the last moment. Their financial position had deteriorated, as almost all their small savings from Du Pont had vanished with the cost of Scotty's special treatment.

Margaret had to maintain the apartment on Delaware Avenue for herself and Jeanie while Norman was paying for the hotel room in Mexico City. It was too much for them to manage.

And finally the doctor was right about the hospital visits. Each time she saw her child, her distress increased. She knew she must go. Norman needed her; he should have a home to go to, not a hotel room. "As soon as you can get a place there, I'll ship our furniture down," she wrote. "We should be doing our work and getting on with life."

Borlaug went to talk with his colleagues at San Jacinto. Wellhausen and Colwell were away in the fields. Harrar was in the process of setting up their first laboratory in the old ministry building. Here they would prepare rust inoculum to determine the disease resistance of their crossbred plants. Borlaug told Harrar he had to find accommodations quickly. With the assistance of Harrar's wife, Georgie, and Wellhausen's wife, Vivien, he soon tracked down a comfortable apartment in the suburb of Lomas, where both the Harrars and the Wellhausens lived, and put down a deposit. "We have our first home in Mexico ready and waiting," he wrote Margaret. "Come as soon as you can." Since they had only a few pieces of furniture of their own in Wilmington, he bought a bed and some chairs and moved in to await her arrival.

When Margaret received the letter, she went to the hospital for a last visit. The doctor gently led her away. She could be sure, he said, that Scotty would be in the best hands. He would see that he had no pain and was as comfortable as possible: "I will keep you informed, and when the end comes I'll take care of everything for you. I think you must leave it there."

It was a terrible wrench for Margaret, but she tried to keep her feelings from her husband. She did not want to add to his own grief. "I think I'll travel down by train," she wrote him. “The scenery will help take my mind off things, and Jeanie will be able to move about. I know it will take the best part of a week once we get clear of here, but I think I’ll enjoy seeing the country.” It would also be cheaper, she knew.

As Borlaug waited for his wife and daughter to arrive, he threw himself into his work as an antidote for his anxieties. It was time to inspect the wheat in the experimental plots at Chapingo, and he took an early morning ride out to the national college compound. He wore an old check shirt, brown twill pants, and sturdy working boots and sat among the Mexican workmen on the bus, smiling but still unable to communicate easily with them. He wondered what they made of him as they talked and laughed with one another. They reminded him of the men in his forestry crew in New England, the workingmen from the slums of Boston, and it irked him that he could not share their talk. He vowed he would learn Spanish quickly. The sooner he got over the language barrier, the sooner he would stop feeling like an alien.

At Chapingo he hopped down from the bus and followed the dusty road past the old mansion that was now the central building of the national agriculture college. The sun gave a rose-colored warmth to the stonework of the tall towers at each end of the building: pinnacles of the past, from which the old landowners, the great families of power and oppression, had kept watch over the workers in their fields. The school—and his experimental plots—were symbols of a new era.

The morning light showed flecks of yellow among the nodding green heads of wheat, the tiny anthers of the first flowering varieties. He knew these early birds. They were the Mexicans, the criollos, strains that would ripen a week or two ahead of the others but that were susceptible to stem rust. The heads of darker green, the later types, were the strains imported from Canada, Australia, Morocco, Kenya, and the United States. They were the varieties collected with built-in resistance to the major killer strains of rust.

There were five miles of rows and many thousands of plants, and from them he hoped eventually to breed the one wheat that came closest to meeting the needs of Mexico. But he faced a more immediate task: to find a wheat that would tide the farmers over until he could breed that ideal variety. It would be folly to offer farmers commercial amounts of experimental seed when he knew that one epidemic of stem rust could wipe out an entire harvest. And even barring such an epidemic, the plants were susceptible to other, less prevalent but equally destructive types of rust, such as leaf and stripe rust, and smut—as well as to root blights and boring insects. There were chemicals to defeat some of these, but they were beyond the means of most Mexican farmers.

He turned to the imported varieties collected in 1943 by Harrar and Stakman for breeding in Mexican conditions. There were some thirty-eight different strains from various parts of the world, all with greater or lesser degrees of rust resistance—but late maturing and with only moderate yielding ability under Mexican conditions. His first task would be to select the best of these rust-resistant types, multiply them rapidly, and get them into commercial production. Meanwhile, he could proceed with the longer-term breeding project: to cross the rust-resistant imports with the high-yield Mexican strains, hopefully combining the best qualities of both in one grain.

He went through the rows of the imports, inspecting the stems for the telltale brown spots, feeling the texture of the leaves and the swelling of the heads, seeking with his eyes and hands for what the plants could tell him. It was a long process of elimination, but Borlaug ended up with four varieties that seemed to have promise.

The first two carried a rust resistance known as the Hope type, which came from a wheat bred in the United States in the 1920s. It was so good for a time that seed firms offered a dollar for every spot of rust found on their plants. Then, as rust had changed its face in the United States, Hope had been crossed with other resistant strains. The two varieties Borlaug chose that had Hope resistance in them, were named Frontera and Supremo, from Texas.

The other two wheats, named Kenya Rojo and Kenya Blanco, had been bred in Africa by an English plant breeder named Burton. Unfortunately Burton had lost all his records in a fire, but it was known that his wheats were a mixture of various European, Australian, and Moroccan varieties.

Frontera, Supremo, and the two Kenya strains—these were to be the cornerstones of Borlaug's wheat program. When they were ready, he would harvest them, plant them for commercial seed, and get them into production as soon as possible. In these four wheats Borlaug knew he had a first-line defense against rust, but it was not enough. Now he had to concentrate on intensified crossbreeding.

Immediately he set to work on this second phase. Among the hand tools in the old adobe hut he used as shelter he found a little three-legged stool with a diaper-shaped canvas seat. He took a pair of fine-nosed tweezers from his pack, filled his shirt pockets with small envelopes, paper clips, pens, and notebook, tied the tweezers around his neck with a piece of string, picked up the stool, and went out to the rows of wheat.

The task of cross-pollinating was intricate and demanding. Many of the wheat plants were nearer to maturity than he had thought and he had to work quickly. The wheat heads already were covered with little flecks of white or yellow from the opening of tiny flowers, which when fertilized would become grains of wheat. It was with these flowers that Borlaug would be working over the next several days.

The wheat flowers, or florets, grow in little bunches, known as spikelets. There can be more than sixty such florets on a wheat head and each floret is bisexual, having both male and female organs. Each has a three-pronged anther, the male stamen, which contains the pollen. This tiny structure bursts when the flower is ready for fertilization. Its thousands of pollen spores, or groins, float out to the filaments of the female section of the floret, the stigma, whose delicate, trailing fronds capture the pollen spores. The pollen grain germinates and develops a germ tube and penetrates to the waiting ovary, where impregnation takes place. Within an hour or two of the anther's bursting, the ovary is usually fertilized.

Since both pollen spore and ovary come from the same source, they carry the same chromosomes, and the new seed that is formed is a faithful re-creation of the parent plant. Thus self-pollination produces succeeding generations of grain without change. Only if the plant breeder deliberately interferes with the fertilization process, or if the plant is accidentally inseminated by a foreign spore, will the resulting grain differ from the plant on which it is grown.

In Borlaug's crossbreeding program the pedigrees had to be clearly defined. Therefore he not only had to prevent his selected plants from fertilizing themselves; he also had to guard against their accidental fertilization from spores floating in the air. His aim was to mate those wheats that gave the best yields with those types that were most resistant to disease, and the technique he used was well established. But it was delicate, time-consuming work, and physically exhausting. Seated on his little folding stool in the hot sun in front of a plant, he would work his way through every head of wheat, using the needle-nosed tweezers to pluck out each stamen from every tiny floret. Not one must be left to self-pollinate; that would defeat his objective. Once each stamen was removed, the bisexual plant was rendered totally female. Now it could be fertilized only by pollen from another wheat plant, and Borlaug had to see to it that fertilization was not accidental.

When each head was emasculated of its stamen, he slipped over it a narrow cylindrical envelope open at both ends. The top of the envelope was then folded over and closed with a paper clip. Each envelope was marked with a code number that would record the lineage of the wheat pedigree. These little envelopes would stay in place until the plants were ready for insemination, a matter of two to five days. Then he would cut the head from a plant he had chosen to be the male parent and remove the paper clip from the top of the envelope covering the waiting female. He would insert the male head into the envelope, twirl its stem between his fingers to release its thousands of pollen spores, withdraw it, and close the protective envelope again.

The most painstaking part of the task, of course, was preparation of the female. His hands had to remain rock-steady; the probing needle-nosed tweezers could easily damage the delicate ovary of the wheat flower. Hour after hour he would drag the triangular stool from plant to plant down the long rows, his back arched and tense as he sat hunched over, his knees up to steady his arms. One hand holding the wheat head, the other grasping the tweezers, his spine bent as he leaned over the heads of wheat, his eyes staring down into the fine structure of the floret. The physical strain was tremendous as his muscles began to cramp, then ache with tension. Yet there was no relief, except when he stood to move the stool from one row to the next.

The mornings were the best part of these long days. The air was cool, the sun low. As the day progressed, the sun's heat grew more fierce and burned the back of his neck. His ears were skinned raw with sunburn and his nose became scorched and swollen. He began wearing a baseball cap so that he could switch its long peak in the direction of the sun—the cap was to become a symbol of the new wheat-breeding era in a dozen lands or more. By noon each day he would be dripping with sweat, and he had to wear a handkerchief around his forehead to keep the perspiration from blinding him. His hands would become wet and sticky, and at times his pen would slip through his weary fingers onto the ground. He would pick it up and curse his clumsiness—and the loss of time.

In the heat of the spring sun the plants were rushing toward maturity. The toil seemed endless. He worked from sunup until dark each day. At night he would go back to the mud-walled hut, open a can of food, heat his coffee over a wood fire, and collapse into his sleeping bag.

At the end of two to five days the females—mostly Mexican varieties that Borlaug had chosen for yield—were ready for pollination. He took the male lines, the rust-resistant imports, and cut and twirled their heads inside the paper envelopes, releasing the pollen over the waiting florets. Then he closed up the envelopes and stood back to let nature take its course. The first generation—known as filial one, or F1—would look like one parent or the other. Not until the third and fourth generation, or F3 and F4, would he see the true characteristics of the new crosses. His work was done for the moment, and he decided to return home for a bath and a change of clothing.

He hitched a ride into the city and went straight to his apartment. He needed rest—and he needed help for the work ahead. He spent the night in the luxury of clean sheets and a bed, and next morning he called Harrar at the Office.

He was only in town for the day, he explained; he had

to get back to Chapingo. But he needed some field hands for a few days. Harrar told him he could borrow some workers from Wellhausen's corn plots near Chapingo. There was a foreman there named Manuel who spoke a little English.

Borlaug did not tell Harrar specifically why he wanted the workers. Ever since his trip to Sonora he had been mulling over an idea that he wanted to put to the test. He had brought back some wheat from the farms in the Yaqui Valley, and now, in May of 1945, he wanted to prepare a field to plant it. If it grew well in the highlands of Chapingo during the summer, he would gather the seed in the fall, take it north again, and plant it in the Yaqui.* The following spring the next generation would be harvested in the Yaqui and would be put into the soil at Chapingo, and so on back and forth. It would mean that he could grow two generations a year instead of one, cutting down his breeding time by half. It would also mean that he could get his four rust-resistant strains into commercial production more quickly—possibly within three years.

It was a simple concept, yet it went against all the traditional precepts of plant breeding. Many Mexicans—and most of his American colleagues—might react adversely to his plan because it had not been tried before: two seasons in one, plus a complete switch from one latitude and altitude to another. But he saw no genetic reason why it should not be done.

Borlaug took a pickup out to Wellhausen's cornfields and found the Mexican foreman in charge. He told him he wanted him and his four men for a couple of days to prepare an acre of land at one end of his experimental station for new plantings of wheat.

* Because of the difference in altitude and temperature, the Chapingo and Yaqui planting seasons were at different times of year.

Manuel had worked many years at the national college at Chapingo before he had joined the Office project. He was astonished at the proposal. His eyes widened, and he spread his fingers in the air. "But Dr. Borlaug, señor, no wheat is planted here in the summer. The rain and heat and the rust will kill it. You will waste time and work, patrón."

Borlaug smiled and patted his shoulder. "Tell you what, Manuel, amigo. You worry about getting that damn land in shape—and I'll worry about the rust. O.K.?"

Manuel's reaction was typical of the fatalism Borlaug and his Office colleagues had met with in Mexico. It was wrong to plant summer wheat in the high country because it hadn't been done before, not because rust would kill it. Rust killed just as effectively in the winter. It was pessimism, deeply ingrained; it was a lack of faith and initiative. And it was a symptom of Mexico's problems. From the highest officials of the Mexican government to the agrónomos and field workers of the national college, there was a pervasive complacency, a reluctance to accept new ideas.

Wherever he went he heard the same old arguments:

"Mexico can never feed itself entirely."

"The climate is wrong; it's too hot; the soils are too poor, the rainfall too low."

"Mexico must buy food from foreign soils and pay for it with other treasures of the earth—its gold and silver and oil."

Borlaug fumed at such nonsense. Mexico needed action and answers, not excuses.

Reluctantly Manuel and the field workers came with a buffalo, and Borlaug put them to work turning the sod and hoeing the soil. It was a wearisome, half-day's job, which could have been done in an hour with a tractor and proper implements. Borlaug made a mental note to talk to Harrar and Wellhausen about equipment.

It was early afternoon when the job was finished. He went back to the shed and took out the seeds he had gathered from the fields of the Yaqui Valley. He sorted them out and planted them in ten different rows in the freshly turned earth, labeling them according to variety and the fields from which they came. Now all he could do was wait. Satisfied that he had made a good beginning, he drove his pickup truck into Mexico City and went to the apartment to prepare for the arrival of Margaret and Jeanie.

The most difficult part of Margaret's journey to Mexico was walking onto the platform in Wilmington and then sitting motionless until the train began to move out of the station. She held little Jeanie on her lap. She was a beautiful child; it was so hard to understand about Scotty. Margaret always believed that Jeanie saved her sanity on that train trip. Her daughter was just old enough to prattle and ask questions; listening to her, Margaret could not hear the wheels beating out the miles that took her away from her baby son. She arrived in Mexico City exhausted from the emotional strain and the long, wearying train journey. But she was joyful to see Norman and he to see her.

Margaret took an immediate liking to the colorful residential suburb of Lomas de Chapultepec. Their apartment was modest but adequate—a living room, small kitchen, and three small bedrooms. When she arrived it was sparsely furnished and had the look of a bachelor's quarters; but she set about to add decorative touches, and it was soon cozy and cheerful.

She became friends with George Harrar and Vivien Wellhausen and several other American women in the community, and together they went on weekly shopping expeditions. She soon learned enough Spanish to shop at the local market, but the language was difficult for her and it was several years before she really mastered it. She never became as fluent as Norman, and neither of them ever achieved the easy proficiency of little Jeanie.

Their reunion ended all too quickly. The wheats at Chapingo—those planted the previous fall—were ready for harvesting. Borlaug had been assigned some of the Mexican workers employed by the Office, but it was obvious that the job was too big to tackle by hand. He discussed the problem of threshing machinery with Harrar but was told that there would be none available for a while. The war in Europe was over, but American industry had not yet caught up with peacetime demand. The best Harrar could do was to arrange for three young agrónomos from the national college to join his crew.

But Borlaug needed machinery and was determined to get it one way or another. Wellhausen was in the same predicament, and together they went out to scour the countryside. Their first find was close to home. In a yard behind the national college they found an old International model TD9 crawler tractor.

Encouraged, they went further afield. On the other side of the city they found a lot where a branch of the ejido bank system dumped repossessed equipment. Two tractors, damaged and broken, were piled in a heap of discarded machinery. Borlaug had a sudden memory of visiting the smithy in Saude with his grandfather, watching the sparks dancing as the blacksmith wreaked some miracle of repair. He turned to Wellhausen. "What do you think, Ed? Suppose we get these up to Chapingo with that other old wreck and see if we can put together one good working model out of the three?"

Wellhausen was hesitant. They were agronomists, not engineers. But why not give it a try?

At Chapingo they faced the real problem. They had none of the tools necessary for dismantling the tractors. They needed skilled hands, engineering talent, metalworkers, mechanics. Where would they find such help?

Manuel, the field foreman, heard them talking and came to their rescue. "Pardon, señors," he said. "I live near Boyeras, and in the next village lives a man named Vicente Guerrero and his cousin José. They are fine metalworkers and mechanics. Also blacksmiths. They have worked for many years in the factory that makes automobile parts in Mexico City. If the señors would wish me to ask them to come and look and see what could be done—"

Borlaug embraced Manuel. Two days later the two mechanics appeared and set to work with skill, determination, and pride. They brought with them a small anvil, and they fashioned a pair of bellows to make a forge. Using hacksaws, hammer, and forge, they improvised other tools, and soon the tractors were dismantled. Pieces of three machines were strewn across the ground in what appeared to Borlaug as chaos. Inside the workshed his seeds and tools were pushed to one side and the floor was littered with piles of screws, nuts, and bolts.

Outside Vicente picked over the hulks of old tractors, and José heated sheets and strips of metal and beat them into various shapes on the anvil. He made cuts from old steel drums, bending and shaping the steel for the cultivation implements, which he was also repairing. Slowly order began to emerge as Vicente patiently assembled the tractor pieces into one working model. Then came the day when they all stood around as he poured oil and fuel into the machine. They held their breath as he cranked the starter. It stuttered into life, and Vicente slipped the gears into place. The big wheels turned slowly, and they cheered. It was their moment of triumph.

Borlaug and Wellhausen could not bear to let these two gifted mechanics go. They saw that they would be needed in the years ahead, and they were persuaded to join the small corps of Mexicans working on the Office project. Borlaug taught them how to use the tractor for cultivating, sowing, and spreading fertilizer.

With the aid of the new tractor, and some old threshing equipment that the two mechanics also put in order, the grains were soon harvested, the pedigrees noted, and the seeds carefully labeled and set aside for the next planting.

Not long after the tractor was put into operation, Borlaug was sent three young trainees from the national school at Chapingo. It was part of the Rockefeller Foundation agreement with the Mexican government that young scientists from the local academies be brought into the program to be trained in the latest theories and techniques of American agronomy. The aim was to improve Mexican agriculture by creating a capable, highly trained corps of scientists able to initiate and direct their own national research programs, and it was part of Borlaug's job to train these young agricultural graduates, or *ingeniero agrónomos,* in the fields. His first three trainees arrived in June 1945. Eager, serious young men wearing white shirts, dark suits, and highly polished shoes, they presented themselves to Borlaug at the entrance to the old adobe hut, where the remnants of the tractor hulks lay in disarray. There was Benjamin Ortega, his intense dark eyes penetrating under straight black brows; Leonel Robles, clean-cut and broad-shouldered; and José Guevera, the youngest-looking, with a hesitant manner and a mouth shaped for quick laughter. They were the first of hundreds of agrónomos whom Borlaug would teach and send out to continue the fight against hunger.

They shook hands all around. In his halting Spanish

Borlaug welcomed them warmly and outlined briefly what their duties would be. Then he dismissed them and told them to report back in the morning.

He was taken aback by their appearance, however. They seemed totally removed from work on the land. Later he complained to Wellhausen: "Damn it, Ed. Won't anyone in that school tell these young people where the real work is done? They come out to the fields in suits and shiny shoes. What do they think this is, a picnic?"

Wellhausen told him how it had been with his first agrónomo trainee, Pepe Rodríguez. He had also insisted on wearing a suit and tie when he went out to work on the corn plots in the Bajío. When Wellhausen complained that this attire might be inappropriate for trudging through fields and across wet ditches, Rodríguez drew himself up to his full height and replied, defiantly: "I can go any place you can go!" And so he had. Stubborn, full of pride, he came out of the fields at the end of the day with his shoes coated with mud and his suit full of burs. His eyes flashing, he said to Wellhausen: "We are used to this sort of thing in Mexico." All the same, he soon bought khaki clothes and sturdy boots for work—but at first he always traveled back and forth in a suit, shirt, and tie.

"Take it easy with them, Norman," Wellhausen advised. "Give them time to see the way of things."

Borlaug felt he had no time to waste on such nonsense. If it was beneath the dignity of these young scientists to work with their hands in the fields, then they had no place in the adventure of the program. Still he had to give them a chance, and he resolved to test their mettle without delay.

When they arrived the following morning in their shirts and suits, he put them to work in one of the wheat fields, loosening the dirt around the roots of the young plants, hoeing, raking, thinning them out—kneeling and bending until they ached. When it was time for the midday meal, Borlaug ushered them back to the adobe hut to eat their sandwiches and drink their coffee in the shade of a tree, along with the other workers. When they started to drift off by themselves, he called them back. It was part of the treatment. They had to learn to work and eat with the field hands, the entire crew.

Each day it was the same. He set the example himself, toiling in the sun alongside the peons; he was always the first to go into the field and the last to leave in the evening. Somehow he had to bridge the gap between the university graduate and the man on the land. It was a wide gulf, he knew, and it had a long tradition in Mexico. To the agrónomo, toil was menial; there was no dignity in sweat. And the opposing view—the resentment and bitterness of the farmer toward the university-educated agrónomo—was equally destructive.

This was brought home to Borlaug forcefully one day when he took his young trainees to one of the biggest farmers on the Bajío to ask permission to plant a strip of seed on his land. The owner was courteous to Borlaug, but when he saw the young agronomists he became vitriolic. His face suffused with anger, and he shouted at them: "Get off my land, you leeches! Don't you think I know you are government people? You suck our blood; you take our taxes and give us nothing in return. You are bureaucratic parasites!" It was tragic to Borlaug's ears. The two sides needed each other so much.

They met with other kinds of resistance and superstition as well. A farmer near Puebla agreed that they could work an acre of his land to demonstrate that wheat could be grown in the summer. He sat idly by, watching them work, until they hitched up his bullock to a steel plow to turn the soil. All his life the farmer had used a wooden plow. When the steel edge bit into his land, he exploded in protest. "No! No! Only wood must be used on my soil. You must not touch the land with metal. It will take away all the warmth, all the life! Without warmth, the soil will be dead."

They grouped around him, trying to reassure him. At length they pacified him, but he grumbled and walked away, pulling his straw hat over his eyes in disapproval.

Borlaug and the young men worked on through the day, fertilizing the land and sowing the seed. The following day they returned to thank the farmer for allowing them to use the land. The acre of land over which they had labored was being trampled under the hooves of a small herd of cows. The farmer shook his fist at the men. His voice trembled with conviction: "You have stolen the heat from my land; only the animals can put it back!"

But if Borlaug met with resistance from the farmers, he was making some headway with his trainees. He finally induced them by example—and by the ruin of their good suits—to wear khakis in the fields, although, like Rodríguez, they would not at first travel to and from the experimental station except in their business suits.

And there were still periodic rebellions among the young scientists to working with their hands. Late in the summer Pepe Rodríguez returned from the University of Minnesota and joined the wheat project. When he saw his young colleagues hoeing in the fields at Chapingo, he asked Borlaug if he could speak with him. He was courteous and polite but pointed out that in Mexico these things are not done. "You lower your status when you work in the fields," he explained. "That is why we have peons. That is their role in life, to work and toil. All you need to do is make the plans, take them to the foreman, and have him carry them out. In Mexico the rule is for the worker to do the work."

Borlaug heard him out patiently. Then he spoke, firmly

and forcefully, clipping out the words so that the other agrónomos could hear:

"And that is exactly why your countrymen are hungry, señor! And that is why we are here in Mexico and not in our homes enjoying our own lives! That is why you have divisions among people who should be working together to grow more food. And all that is exactly why things must change. You must work with the farmer and he must respect you! And you must work with the plants, you must know what you are growing. You cannot delegate it to a foreman and expect him to give you a scientific analysis. If you leave the work of your experiment to the foreman, how will you trust your results?"

He hammered his message home without pause.

"I know how you feel about working in the fields, but this is the way you have to work if your science is to be effective. If you want to grow more food for Mexico, then you have to get to know the plants and soil; and the only way to get to know them is to work with them, live with them, listen to them. If you listen properly, they will tell you many things. They will talk to you as they talk to me."

Gradually the young scientists came to share his belief. Through his words—and particularly through his example—he instilled in them a pride in their work, a dignity in their sweat.

As the summer wore on, Borlaug kept a close eye on the small field he had planted with seed from the Yaqui Valley. Contrary to the predictions of the foreman, Manuel, the wheat was thriving, and the sheen of green, waving stalks was beautiful to his eye. Here was living proof that he could grow two seasons of plants in a single year.

The idea of breeding two generations in one year was not entirely new. Dr. Stakman, at Minnesota, had discussed it during one of his "coffee klatsch" sessions with some of his graduate students. He had actually asked for money for heated greenhouses so that he could grow wheat all year round. But it was an approach that traditional agronomists then regarded as heresy. Seed needed a period of adjustment, they argued, a rest between harvest and planting. There had to be a sufficient break in time for it to store the necessary energy that would result in germination and the sprouting of new plant life.

Borlaug had heard the argument often. In his sophomore year at Minnesota Dr. H. K. Hayes had drummed this dictum into him as fundamental. But now he believed the concept was wrong. The field before him was proof that he could take seed from a winter harvest and grow it in a different climate as a summer crop. Now he planned to take it the next logical step and return it to the Yaqui Valley in the coming fall.

Borlaug had discussed his plan with Wellhausen. Wellhausen, he knew, had the ear of the project director, Harrar, and was his closest confidant. It was in the breath of this friendship that he indicated to Borlaug one evening how the wind was blowing against his idea.

Doubling a wheat-breeding season had more to it than planning. It doubled the work, doubled the number of time-consuming soil tests, and, above all, doubled the strain on the project's limited resources.

But Borlaug was determined. "Don't try to discourage me, Ed. I know how much work is involved. Don't tell me what can't be done. Tell me what needs to be done—and let me do it. There's one single factor that makes the Yaqui effort worth a try, and that's rust. Breeding two generations a year means beating rust. If I can lick that problem by working in Sonora, then we've won a victory. To hell with the extra work and strain. It's got to be done, and I believe I can do it."

He had been given the first indication of disapproval and he had given his reply. He pressed ahead with the selection of seed for replanting in the Yaqui and made plans to fly north.

Seven

Borlaug packed the seed from the first generation of Sonora wheats harvested at Chapingo, and seed from the other strains bred in the Chapingo station—the rust-resistant lines of Supremo, Frontera, Kenya Rojo, and Kenya Blanco. Then he boarded the rickety six-passenger plane that took him once again to the airstrip near Ciudad Obregón.

It was early fall, and the sun cast long, corrugated shadows across the wooden sidewalks of Obregón's main street. After buying a few supplies and a cheap straw hat in the crowded marketplace, Borlaug hitched a ride out of town. The driver let him off at a crossroads. From there he had to walk two miles of flinty road to the research station. His leather boots stirred up small clouds of dust that soon coated him with an ocher-gray film. The powdery coating, the bedroll and pack on his back, and the frying pan and coffee pot hanging from his belt made him look more like a prospector—or an archeologist—than a biologist.

As he passed between the two stone pillars at the entrance to the station, Borlaug could see that little had changed since his last visit. The old Mexican caretaker apparently had gone for the day—if indeed he had been there at all. The place was just as Borlaug remembered it, a neglected, deteriorating shambles, overgrown with weeds, deserted except for the silent buzzards in the trees. Borlaug knew better than to be optimistic about the possibility of improvement in the future. He had been to Mexico City to see the director superintendent responsible for the station, Inginero Ricardo León Mauzo. Mauzo had given Borlaug permission to live and work on the site and had made arrangements to prepare the land for planting. But that was all he was prepared to do.

When he arrived at the research station Borlaug heated a can of stew, brewed some coffee, and brooded over all that needed to be done. His first task was to make the place livable, and he started work early in the morning, stripping terracotta tiles from one of the old buildings and patching the roof of the small cottage he had chosen to occupy. He built a Mexican-style hearth and oven from rocks and stones gathered in the field and fashioned a sturdy cot of poles and tightly stretched canvas. Then, his few possessions laid out on a wooden table, the scientist opened his notebook and began to outline a plan for his work in the fields.

Late that night, as he lay in his sleeping bag on his cot, Borlaug thought about the plan he had worked out. It was too big a job for one man working alone with hand tools. If he was going to get the seed planted promptly, he would have to find equipment and a means of power—animal or machine. Wellhausen was right: Borlaug was going to have to muster twice the financial and human resources available to him if he was to succeed—and he would have to double his own labor and bear with twice a normal day's headaches. But if he could defeat rust before rust defeated Mexico, the investment, however great, would be worthwhile.

His goal was to transform the neglected 250-acre station into an orderly experimental area, a place where Yaqui Valley farmers could come to see what they could achieve on their own farms. By exposing them to the products of successful experiments, Borlaug hoped to change their ingrained attitudes and outmoded techniques. But at this stage, before he had anything at all to show them, Borlaug knew that getting their help would be difficult.

It didn't take him long to discover how difficult. The next day he set out along the Yaqui Valley road to solicit the help of local farmers. Invariably the scientist was received politely; he was never run off any man's land. Yet virtually every farmer he met revealed his suspicion and apprehension—and sometimes a scarcely concealed hostility for Borlaug's gringo origins. For the most part the Yaqui farmers were men of property and substance, men with good lands, machines, and hired help. But it seemed to Borlaug that the larger their holdings, the more machinery and wealth they possessed, the more guarded and grudging they were.

"Could you lend me a small tractor for a day or two?" he would ask, explaining that it would be used to plant a new kind of wheat that would grow well in their valley. "I need some implements to prepare the soil at the station. I need plows, disks. Can you help me?"

Their eyes would widen in astonishment. Here was this gringo, in work clothes and boots, asking in the most atrocious Spanish for their valuable equipment. Why was he not in white shirt and tie like an agrónomo? He became a subject of gossip along the valley. Who sent him here? He must be crazy. If a man gave him expensive machinery, what would happen to it? It might disappear.

Borlaug then went to the smaller farmers along the road to see if he could borrow mules, but the response was the same. He got tired of begging, but he continued his rounds until the daylight faded.

It was late in the evening when Borlaug returned to the station, angry and exasperated. He warmed his supper over the fire and pondered the situation. The resistance here was as strong as that of the poverty-stricken people of the Bajío—and yet these farmers were not ignorant. They ought to know better. They were blind idiots, fools consumed with distrust. He couldn't ask for their help any more; he would have to go ahead without them, however long it might take.

In the loneliness of the Yaqui night Borlaug's anger gave way to recollection. He reviewed the events of the past months. Maybe he had taken on too much. Perhaps it was unrealistic to expect to transform a whole rural industry in a nation where change itself was resisted—and resented when it came from without. But he couldn't think such thoughts for long: they made him despondent.

Of course, he could have stayed in Wilmington; he could be going home each evening for cocktails, a hot shower, a leisurely, filling dinner. Or he might have found a comfortable position in a good university. But here he was, more than 1,000 miles from his family, and here was Margaret, stuck in a small apartment in a foreign land, eking out his pay to keep poor Scotty in the special hospital ward. Suddenly he sat up, shouting into the empty night: "What the hell! What in heaven's name am I doing here?"

He could pull out. The Yaqui Valley project was still nothing but an idea—an idea his colleagues opposed, an idea the farmers distrusted. He could pack his bedroll next morning and hike back to Ciudad Obregón. Who would care? Who would blame him?

But no. All his life he would have to live with the knowledge that he had run away because the job was too hard. He would rather stay and fight. He closed his eyes and slept.

In the bright light of early morning the buzzards, scratching at the soil, and the boat-tailed grackles, flew away at his approach. A few strands of candy-floss clouds marred the pale sky, and a light breeze was blowing out from the gulf, carrying a fragrance of rich soil and clean ocean.

He went to the sheds and rummaged among the old equipment. He found wooden handles, pick heads, broken hoes, and rakes and started repairing them. Searching further for hammers and nails in a dark corner, he found an old wooden hand cultivator, unbroken and with the harness attached. He took it out on the flagstone courtyard and was washing the encrusted dirt from the blade when the Mexican caretaker arrived. "Buenos días, señor," he called, smiling through brown teeth.

Borlaug put him to work. They carried the plow to the field, and he put the caretaker behind it. Then, fitting the rope harness over his own shoulders, he started to cut wavering furrows through the earth. It was a cool morning, and although the ground was easy to break, friable and loose, within a quarter of an hour Borlaug was running with sweat. His back ached with the tugging; the rope cut through his shirt into his flesh, and he began to stumble. But he would not stop. Once he looked up and saw two men watching him from the road. One of them held a horse by the bridle. He could use that horse, he told himself, but he would ask nothing from them. He thought of the peons he had seen in the high mesa country. Now he knew their drudgery.

Nearly exhausted after two hours, he stopped. While he rested under a tree, he sent the caretaker back to the sheds for hand tools. All through the afternoon he worked, and by evening he had planted a dozen rows of seeds. He followed the same pattern the next day. On the third day he was raking the soil over a test bed in which he had sown some of his new wheat crosses when he saw a farmer from over the hill walking across the field toward him. He was wearing his best suit, and Borlaug realized it was Sunday.

"Buenos días, señor. I am Aureliano Campoy. I live there, you know?" He pointed a finger back over the hill. Borlaug recognized him as one of the farmers he had approached for help. Campoy had shown more interest than the others but was skeptical of this government man who came like a peasant. He was one of the few farmers who spoke some English. It was heavily accented, but understandable.

"I must ask you. I see you working here like a peon. Why do you work on the Lord's day? What is it you do here—a man who turns himself to work like a mule?"

Borlaug tried to explain that he had a deadline to meet, that he must get the seed in in time to harvest for replanting back south on the Bajío. "I don't think he believed me," he wrote Margaret that night. "But I know he was very curious. I had seen him watching me work from across the road. He had to have an explanation; so I told him we were making new wheats—and glory be!—he will let me have his small tractor and some implements next weekend."

Although Campoy's equipment speeded up the work on the Yaqui station considerably, Borlaug could use it only on weekends. The rest of the time he used hand tools, working straight through the daylight hours until he returned each night, stiff and aching, to his little room. One night, as he was cooking his supper over a wood fire, he suddenly became aware that he was talking to himself. It occurred to him that apart from the few words exchanged with Campoy, his own was the only English he had heard in many days.

Depression and doubt began to discolor Borlaug's days. When he caught them closing in, he tried to head them off and to analyze his moods objectively. Why was he suffering more now than before? Was it the lack of contact with other scientists, other Americans, other people with whom he had something in common? Or was it just the solitude? And if it was, why hadn't solitude closed in on him so oppressively in the Salmon River Mountain Wilderness? Perhaps it was the overflow of new pressures that was getting him down, or the grief of Scotty's tragedy, or the separation from Margaret and Jeanie. Perhaps it was the endless toil without conspicuous results. Probably it was an accumulation of catalysts, Borlaug decided—and there was nothing, short of quitting, that he could do to alter his circumstances and lift his spirits.

There was nothing he could do of his own accord, but a salve to his moodiness did come along, and it came in an old Chevrolet on a Saturday morning. Borlaug was working by the station with the tractor, cultivating his five acres of antirust wheats, when he looked up to see a jalopy bumping down the road from the direction of Campoy's farm. He had never seen the vehicle before and was surprised when it jerked to a halt in front of the station, the horn blaring to get his attention. He cut the tractor engine and watched in fascination as the driver emerged from behind the wheel. She freed herself with difficulty, shifting her enormous bulk through the door until she stood beside the car. Borlaug guessed she must have topped 300 pounds. She moved toward him, waved a massive arm, and called in a bell-like American-accented voice, full of warmth and friendliness: "Hey there, sonny! You must be the crazy American my foreman has been telling me about. Come on over here and say hello!"

He hopped down off the tractor and crossed the field, grinning delightedly. Putting his hand in her pudgy fingers, he said: "Howdy, I'm Norman Borlaug." Before she could respond he was already telling her about what he was doing there, the words spilling out almost too fast for comprehension.

She looked at him kindly, and when he paused she said: "I'm Mrs. Jones. Everyone knows me here. I live across the canal there, next to that old curmudgeon Campoy—and he don't talk much, other than what I've taught him." Borlaug liked her directness. "I'm not prying," she continued, "but I hear how you haven't got any transportation, and I guess you've got to eat same as everyone else. I go into town every Saturday for groceries and things. Maybe you'd like to ride along and get some things you need."

He went with her into the town—and many Saturdays in the months to come. Mrs. Jones almost mothered him. She took to calling for him on Sundays and taking him back to her farm for "coffee and cake," which usually turned into a full-sized meal. She did his washing and mending and listened as he talked of his work and plans. And bit by bit she told him her own story.

Mrs. Jones had come to the Yaqui as a young bride from Connecticut. She was eighteen and her husband twenty-one; they had been drawn by an advertisement in a California magazine. The Richardson Land Development Company was offering land in the "fabulous, booming Yaqui Valley," where the company owned the water rights of the river and where a canal was being built for irrigation. The accompanying photograph showed lush, irrigated fields, and the Joneses and many other families had swallowed the bait. But the canal turned out to be nothing more than a ditch, and much of the land was useless. Most of the families went back home, but the Joneses decided to stay and try to carve out a life for themselves.

"We started with a dry ditch, a few tools, and not a dollar left in the bank," said Mrs. Jones. "But we built a home near the Yaqui River and captured flood waters in summer with a system of dikes so as to flood the land—and sow it in winter to wheat. We put in wheat and flax and cotton; we worked and we survived."

She told him stories of bandits, Indian wars, battles and skirmishes between revolutionary factions. "I know places along this valley where the wheat grows on land that's been soaked with blood. And we brought up two daughters amidst all that."

Borlaug never met her daughters. They had been sent back to school in the United States, where they married and settled down. When her husband died in 1930 at the age of fifty-five, Mrs. Jones resisted the pressures to pull up roots and go back to America. "I just couldn't do that. This place had been my home for most of my married life. My husband gave his heart to this place—which ain't so bad now the water's here. We fought and slaved for this land. And now he sleeps here. So I will too."

Mrs. Jones was a mine of information and a tower of strength to Borlaug in the coming years as he struggled in the Yaqui Valley. She was his friend, counselor, and refuge from despondency. Coffee and cake with her always renewed his faith that what he was trying to achieve could be done. "I've got no way to say what I owe you," he said to her one evening, "except that if you had not been so good to me I might have gone plumb crazy down there on that station."

They both laughed, but it was true that this strong, wise woman turned the Yaqui from an alien land into a friendly place.

One evening when the seeding was almost finished Borlaug borrowed her Chevrolet and drove the miles across the valley to Roberto Maurer's home. The Maurers did not know that he was in the valley and were delighted to see him. Teresa cooked a special meal as he talked with Roberto about his wheat breeding.

After dinner Roberto told him about a cooperative credit union the farmers were forming so they could borrow money in times of trouble and finance expansions and improvements on their lands. As he talked, the name of Don Rodolfo Calles came up again and again. Borlaug recognized him as the man who had set up the research station years before.

"He is the driving force in this community," Roberto told him. "He has been working for many years in a quiet way for a better life for these people; and where he leads, most of them will follow."

It was early December when Borlaug flew back to Mexico City. When he reached the apartment in Lomas, he could read the sad news in Margaret's face. She kissed him and told him—Scotty was dead. They held each other, close in their grief, and in their relief. Margaret drew what slim consolation there was in the baby's death. "I prayed he would go without pain, in his sleep. He never had a hope, poor little thing."

She said she had wanted with every fiber in her body to rush to the airport, to fly north to see Scotty for the last time, to be at his funeral. But the doctor had discouraged her. It would be best for her to keep the memory as it was. It would be kinder for her. She realized the wisdom of his advice. Also, she did not have the money for the airfare. There would be the cost of the funeral and the final accounting of the hospital.

Borlaug consoled her. "It will get better with time." But they both knew the scar would last for life. It had been a great strain, particularly on Margaret, and Borlaug decided to stay with her a few days before going back to duty at the Office.

When he returned to the project headquarters at San Jacinto, Borlaug found it busier and more crowded than ever. Harrar had doubled the staff, adding more than a dozen new Mexican graduates and several young scientists from the United States. Somehow—although more than twenty people were now using the three rooms and the new laboratory—Harrar had managed to impose greater orderliness on the quarters. He had also persuaded the Mexican ministry to erect a greenhouse on the premises, and his library was expanding by leaps and bounds. One new staff member, Dr. Dorothy Parker, a botanist and biologist, had a flair for bibliography that she applied to give order and greater usefulness to Harrar's collection. During her long tenure with the Rockefeller project she amassed a collection of data that would be consulted by scientists and scientific institutions around the world.

Dr. Lewis M. Roberts had joined the corn-breeding program. Colwell, the soil scientist, had returned to the States. A tall, friendly Texan with a doctorate from Yale, Roberts had worked with Mangelsdorf and was held in high esteem in the scientific community. Other new recruits from the United States included Dr. John J. McKelvey, an entomologist with international experience, who had studied under Harrar in Virginia; and Joseph Rupert, a young graduate from the University of Minnesota, who would be joining them soon as a research assistant on Borlaug's wheat program.

The Office was expanding on another front. Harrar had made plans to erect a building on the Chapingo site to house research laboratories and to accommodate an office staff for an expanded experimental program that he would base there. "When we're through out there at Chapingo, that place will never be the same again," he told Borlaug.

The force that was assembled late in 1945 was a strong and dedicated team. Their skills included botany, plant breeding and pathology, biology, entomology, genetics, agronomy, and soil science. Their efforts were concentrated on improving corn, wheat, and bean crops, enriching the soils, and rejuvenating Mexican agriculture—and most important of all, on training a corps of young graduates who would eventually take the reins of their own national research. But while it was a strong force, it represented only a modest investment in terms of the millions of dollars the Rockefeller Foundation was to pour into the program in the years to come.

This kind of commitment was not new to the Rockefeller institution. The foundation had been working on a global scale for thirty years in the field of public health, fighting plagues, disease, and ignorance and using its vast sums to combat human misery. Founded by John D. Rockefeller, Sr., in the first decade of the century, it had its roots in the same desire that had motivated other great philanthropic institutions through the years—the desire to use great wealth for great good.

The history of the Rockefeller Foundation is long and complex. After some difficult years of independent operation it was incorporated by the New York State legislature and primed with 72,000 shares of the Standard Oil Company of New Jersey. By 1914 Rockefeller had settled a total of $100 million on the foundation; upon his wife's death, the settlement was increased to more than $250 million when he incorporated funds from her memorial trust. During the years of the Depression income fell, but afterward the inflow of funds steadily increased.

Each year tens of millions of dollars were channeled into a fight against human misery. The common cold was studied; hookworm and yellow fever epidemics were suppressed. Virus diseases were tracked down and unraveled, and new vaccines were produced—some by the millions of shots. At the end of World War II many cities in Europe were saved from typhus epidemics by the Rockefeller Foundation's powder spray guns, using DDT against disease-carrying lice. Among the individual scientists assisted by Rockefeller funds was the Oxford professor Howard Walter Florey, who developed penicillin for everyday use. By the early 1970s the foundation had dispensed close to a billion dollars to promote the well-being of mankind.

But the team that went to Mexico in the early 1940s represented the Rockefeller Foundation's first long-range commitment to fighting hunger on a foreign stage. At that time world involvement could not be foreseen. There was not yet a Food and Agricultural Organization, not even a United Nations. The wider horizons of the hunger fight —in Asia and South America—were in the mists of the future.

The war and its aftermath curtailed the Mexican program, however, and by the end of 1945 the Rockefeller commitment was still modest. Supplies, equipment, transportation, machinery had not yet begun to flow with peacetime ease. Harrar had to husband his resources—men and materials—and he demanded the same hard work and self-sacrifice from his colleagues that he himself gave. In return he allowed his staff the greatest possible freedom to work in whatever way suited them best. It was the exercise of this freedom, however, that was ultimately to bring Borlaug and the project director into conflict.

Aware of Harrar's tacit disapproval of his Yaqui Valley scheme, Borlaug concentrated his efforts for the time being on the new wheats at the Chapingo station in the Bajío. But the Yaqui was never far from his thoughts. He made several forays into the Bajío, usually with two or three Mexican graduates, and searched for farms where he could test-grow strips of the new varieties of wheat he would bring back from the Yaqui Valley station after the harvest. Early in 1946 a new agrónomo, Teodoro Enciso, was assigned to his team. Enciso was an enterprising graduate whose English was fair, and Borlaug decided to make use of his bilingual ability when he went back north to Sonora.

In March of that year Joe Rupert also joined them and was to become one of Borlaug's closest and strongest allies. Borlaug and Rupert had a great deal in common—even to their physical appearance. They looked so much alike that they were often mistaken for each other, and some officials in the Mexican ministry swore these two Americans must be brothers. There were other similarities: both came from Iowa, both were products of the University of Minnesota, and both had studied forestry. Though they had not known each other at school, they had had many of the same instructors, and Rupert had followed just a year behind Borlaug. He had graduated from Minnesota, then took a master's degree in plant pathology at Michigan State University, and now planned to write a thesis for his Ph.D. for presentation at West Virginia University. Both had fallen under the guiding hand of Dr. Stakman.

The two men had a close affinity in their view of their work and its meaning. Rupert had been drafted into the army during the war and had gone to North Africa with Eisenhower's forces. He had come back with a fiery anger against waste and killing and a determination to do something worthwhile. He asked Stakman and Harrar for a post with the Mexican project and was quickly assigned to Borlaug. Companionable, quick to grin and sympathize, he was almost as vocal as Borlaug in his criticism of officialdom and folly.

Rupert had spent his first weeks in a hotel room in Mexico City, just as Borlaug had, but as their friendship ripened Margaret decided to offer him their spare bedroom. From then on Joe Rupert was a member of the Borlaug household.

It was Rupert who first brought Richard Spurlock to Borlaug's attention. Spurlock was a farm manager who had once farmed in the United States and now managed his own rented land near Toluca, as well as a number of holdings for absentee American owners in the Toluca Valley and in parts of the Bajío. Tough, hard-working, and professional, Dick Spurlock was built like a stunted oak tree —short and thick, with powerful shoulders and hands. He had a gimlet eye and a quick mind for innovation.

Rupert had run across him on one of his trips to the Bajío, looking for farmers to cooperate with the wheat-growing project. Spurlock managed two large properties near Toluca—Rancho El Carmen and Rancho El Cerrillo —and in the winter he had hundreds of acres lying idle. He also had hundreds of peons on his payroll, barns for storage, and modern equipment. He quickly grasped the value of Borlaug's ideas and offered to cooperate by using his farms to multiply the antirust wheats on a commercial scale. It was a great leap forward in production potential.

Borlaug went to Spurlock with another problem: he was going to need machinery for the coming harvest in the Yaqui Valley. Spurlock could not spare any of his own, but he went to his library and found a scale drawing of a threshing machine in an American magazine. A good mechanic might be able to make something of it.

Borlaug took the drawing to Vicente Guerrero. Could such a machine be built with the forge and anvil out at the adobe hut? Vicente studied the plans. He could build the framework, he was sure, adapt the bearings and cut the threshing flails from pliant metal strips. But there was one other problem. "Such a machine, señor Borlaug, could not be turned by hand. It must have power, and I cannot build an engine."

"You and your cousin build that machine as fast as you can, amigo," said Borlaug. "I will find you an engine!"

He spent the next few days with Rupert touring the junkyards and back-street shops of Mexico City until he found a small gasoline engine. It would have to be fitted with a drive wheel to turn the threshing machine, and that again called for Vicente's special skill. It took weeks of work and innovation to fashion the machine, alter and adjust the parts, and fit it with a working engine, but it was ready for operation by the time Borlaug had to leave for the harvest in the Yaqui Valley.

Difficulty in air-freighting the machine, other equipment, and supplies made him decide to go overland for the first time. Arrangements would have to be made through Harrar at the Office, and this led to the first open opposition to his plans for work in Sonora.

Dr. Stakman was visiting the premises as a member of the Rockefeller Foundation advisory committee when Borlaug approached Harrar. He saw Harrar's face grow serious with the same poker look he adopted when dealing with difficult Mexican officials. Harrar shook his head; it was far too risky to go overland on the west coast road. There was country there to match parts of the Grand Canyon. But why did he want to go north at all?

"Listen, Norman, it makes no sense to risk a round trip of 2,500 miles through that country. You could lose everything—and for what? The guts of our problem is here, in the poverty areas. We've got to win our fight right here, in the breadbasket. You've got to get that clear!"

Borlaug's chin thrust out, his brows knitted; he stood facing Harrar, bent forward slightly, his hands held away from his body. Stakman was reminded of a night on the wrestling mat at Minnesota when a young graduate student fought a bigger, stronger opponent. Here was another confrontation, a clash of wills between two strong-minded men.

"I'm going north because there's potential there. What we do there will soon be followed in the Bajío. I'm going there because I think I can lick stem rust up there and increase production rapidly. And I'm going because I think it's right!"

Harrar shook his head, his face remained stern. Borlaug

lowered his head a little more. Then suddenly he said: "All right, Dutch, we'll settle this now. I'll wrestle you for it!"

Stakman saw a ghost of a smile on Harrar's lips, as he replied: "Like hell you will! Why should I give you the advantage? You want to settle things that way, I'll run you a quarter-mile footrace."

Borlaug did not move. "I'm going up there to work, Dutch."

Harrar shrugged and half turned, looking at Stakman. "Have it your way—for now. But you'd better not try to make it by the coast road. Take the northern route, up through El Paso, across Arizona to Douglas, then down through Nogales to Sonora. It's longer—but safer and surer."

The issue was not settled, just postponed. Authority for his project in the Yaqui Valley still hung fire.

Borlaug decided to take the route Harrar recommended. A few days later he packed the truck and pickup with the help of his Mexican trainee, Teodoro Enciso, and Joe Rupert. The homemade threshing machine was strapped in with their personal gear, next to a sixty-gallon drum of gasoline. It all looked precarious. Margaret and Jeanie were there to see them off. They said good-by, and Borlaug gave his assistants last-minute instructions on the care of the wheat crops at Chapingo. Then he kissed Margaret and waved from behind the wheel as the old pickup went bumping down the street from San Jacinto.

Margaret watched him disappear and tried to hide her disappointment. How many times had she seen him go like that? And how often would this scene be repeated in the years ahead?

Three days of hard driving over rough roads took them to Ciudad Juárez, across the U.S. border from El Paso. They spent the night at a small hotel, and next morning Borlaug drove up to where the tin customs hut stood in the sunshine. He was anxious to arrange passage across the border for the pickup truck and its contents so he could reenter Mexico through Douglas, Arizona. He had been told that there was an agreement to allow such journeys. He had not been told that everything depended on the whim of the United States customs officer.

He walked into the hut in good humor, expecting soon to be on his way. Hours later he emerged, red-faced with anger, sweating, and frustrated. He had met the wrong man at the wrong time. The U.S. customs officer was stolid, unyielding, emotionless. Borlaug had contained his irritation—at first. He went over the reasons for his journey through Arizona but met with a wall of obstinate officialdom. Agreement or no agreement, said the official, if Borlaug wanted to take goods into the United States, he would have to pay duty. It made no difference that he planned to take them back into Mexico.

Borlaug began to boil. He asked if he could make a collect call to New York to have the Rockefeller Foundation confirm his travel arrangements. The customs officer only grew more hostile. "Look," he growled, "you kin call the goddamned White House if you want to—ain't going to do you any good. You want to cross with these goods, that truck, you do it in bond, or you pay duty. That's just how it is."

There was nothing more to say. Late that afternoon Borlaug contacted an El Paso bonding company to ship his precious equipment and threshing machine to Nogales. He pledged the duty on the pickup truck through a local bank, should the vehicle not be brought back into Mexico.

Next morning the bonding company truck got the machine and goods through. It took Borlaug another four hours to get the pickup cleared on the duty pledge, and then it was too late to catch up with the truck. They drove across Arizona, stopping now and then to eat but not to sleep. When they reached the Mexican border town of Nogales the bonding company office was closed, and when it opened next morning the clerk looked blank. He knew nothing of Borlaug or his goods. Establishing the facts took several more hours, long-distance telephone calls, and money. The customs man at El Paso had not given the bonding company permission to bring the goods back into Mexico, and the truck had continued its run to Los Angeles. It would now take some days to ship the consignment back to Nogales.

Borlaug, Rupert, and Enciso found a small hotel, walked about the border town, and seethed at the delay. Precious days torn away from the harvest season. On the third day the truck arrived and the goods were loaded into the pickup truck. They headed south, driving through the night. Next day they reached Hermosillo, the capital of Sonora, and pressed on across to the Yaqui Valley and the station. It was almost two weeks since they had left San Jacinto.

Borlaug's anger vanished at the sight of the wheat. Row after row were rushing toward maturity, and at first search he could find no traces of rust attack. It looked as though the Hope and Kenya types were holding out against stem rust. Moreover, his new breeding lines derived from crossing Mentana and Marroqui, two Mexican strains, with Hope and Kenya types, also showed up healthy and luxuriant, untroubled by rust. He felt light-hearted.

Now that he had transportation he wasted no time in contacting Roberto Maurer and his wife, Teresa. Along the way he stopped to examine fields of mixed Mexican wheats. Everywhere he saw evidence of stem rust—it was not epidemic rust, but it was there, unlike his own.

When he got back to the station he and Enciso unloaded the threshing machine and other supplies. Then he began preparing some of the wheat strains for crossbreeding. This time he was seeking a fresh arrangement of genes and chromosomes that would give the new plants resistance to diseases other than the major races of stem rust. He also hoped to combine the resistance of the Hope wheat and the Kenya variety with the yielding capacity of the Mexican wheat.

As he moved from plant to plant on his three-legged stool, with his sharp-nosed tweezers, he talked to Enciso about the work. "It is a long shot. Maybe in only one crossed floret will we get the two sets of genes that make the combination we want. But once it appears we have to be clever enough to identify that plant—to recognize it for what it is, select it out, grow its seed, and regrow it until we have a wheat with the special characteristics we need. Then we shall have a gold nugget for Mexico."

As the spring harvest approached, he began to give more thought to where he might test-grow the new seed next fall. Roberto Maurer had promised to plant some test strips, but Borlaug wanted to get more of the local farmers involved. One evening after a long day in the fields he went to see Mrs. Jones. She knew everyone in the valley and would be able to tell him who the most influential men were.

"I figure this way," he explained. "You grow beautiful experimental fields at the station that the local farmers can see and admire. But they don't relate them to their farms, to their everyday work. I want them to see these wheats on the land of the farmer next door. Then they think: 'If he can grow such crops, so can I.' Once they believe they can do it, we generate competition—and production. But I need the best farmers, men to lead the way. Who are they?"

He knew about Don Rodolfo Calles, he said. But who were the others?

Well, there were Don Jorge Parada, Don Eduardo Vargas, Inginero Rafael Fierros, the Esquer family, and people from the Obregón and Gallagos families. "They are all big, important farmers," she said. "But you ain't going to get anywhere at all, sonny, till you get on the side of Don Rodolfo. He frowns on a thing, it's finished. He thinks something is good for the valley, everyone goes along with him. But he's a tough one. It won't be easy. Remember, he built an experimental station here in the valley when he was governor—and then saw it mismanaged and disintegrate."

Don Rodolfo Elías Calles had been active in the federal government for many years, first as minister of communications and later as governor of the state of Sonora. He had since retired from politics, but as the son of former president Plutarco Elías Calles, founder of the ruling Partido Nacional Revolucionario, he had influence with many important people in Mexico.

"I hear he likes the bullfight," Borlaug said. "I think I'll go grab him by the horns and wrestle with him."

"You be careful, sonny," replied Mrs. Jones, "that he don't pull out all your tail feathers!"

Four days later Borlaug and Enciso arrived unannounced at the Calles farm. It was a rambling hacienda with white painted rails and horses grazing in a green paddock, and it spoke of money, pride, and care in management. Don Rodolfo was at home; he ushered them into a spacious living room, looking at Borlaug with a slightly puzzled expression in his brown eyes. Six feet tall, erect of carriage, and distinguished, with a broad, strong face and pronounced cheekbones, he was a man of rank and looked the part. Enciso was a little overawed, but Don Rodolfo was courteous; he offered them coffee and asked if they were well.

Their host made polite conversation about the weather as Enciso translated. Borlaug sipped his coffee, waiting for a chance to talk about the wheat research at the station, but it was obvious that Don Rodolfo would talk of anything but that. When Borlaug tried to come to the point, his host commented on the hope brought by the United Nations to the world's food problems. He made it clear that he considered this a courtesy call—nothing more. After coffee he stood up, waiting, his face wreathed in a smile. "You must come again," he said, ushering them to the door.

It was purely a formality, but Borlaug did not want to let the opportunity pass. "Thank you for that invitation, Don Rodolfo," he said. "I would be happy to come again."

Mrs. Jones was right. It was not going to be easy to win an alliance with Don Rodolfo.

He called on Campoy and paid visits to other prominent farmers. But there was no breaking the ice with them yet. He was still the crazy guy down at the station. He continued to work hard on his breeding program, and his fields were studded with hundreds of little white envelopes. Now there would be several weeks of waiting, until the seeds filled and the heads of the antirust wheats turned gold.

Borlaug used the time to patch up more of the dilapidated buildings and clear new fields and irrigation ditches. One day he walked out to where the water in the main canal tumbled over a concrete weir, splashing its way into the ditches and through the furrows to feed the commercial wheat stands. More than 100,000 acres along that part of Sonora could tap the water to nurture grain and other crops on what had once been semiarid desert.

He sat down to rest under an old, bent willow tree, and his mind began to wander back in time. Water had always been vital to man. When men were hunters and gatherers, it was vital only as a drink, but sometime in the neolithic period it had taken on more crucial value as a feeder of crop plants. That was when the women of the wandering tribes (almost certainly it had been the women, he believed) had domesticated the first wild plants. They began to cultivate corn, wheat, barley, oats, rue, and rice—the crops that would feed man throughout history. Helped by nature, they managed biological changes and genetic shifts without knowing such things existed. It was a great conquest over primitivism. It changed tribal territory into soil. The hunter turned farmer and man was on his way to community life.

Famine and disease meant much to early man. There was fierce competition in nature for man's daily bread: locusts, aphids, weevils, rodents, borers, rust, mildew, and associated pestilence all competed for the food. In this valley there had always been land and water. Now there was something new—a biological defense against disease. Yet it was only a step. How far could modern science go in developing a shield against the entire range of rust diseases in wheat?

Suddenly it occurred to him that rice was immune to rust attack. It had its pests, virus enemies, organic predators—but it had stood inviolate to rust through all of history. Might there be some way to cross wheat with rice, to breed the best traits of both into a single plant? Crosses had been made between wheat and rye, but they produced infertile plants. It was the same genetic barrier that made mules unable to breed. Was there some way to cross this barrier with wheat? By mating and crossing thousands of lines of wheat, might he find one that would cross with rice to produce a fertile—and totally rust-resistant—plant? It was only a thought, but it was to have vast importance in the years ahead.

They threshed the small harvest at the Yaqui station, bagged the seed, and made preparations to leave. They decided to take the coast road, regardless of the dangers. It was a long journey, through Mazatlán, across trackless cactus desert, ravines, and fordable rivers, and over the Sierra Madre. But there would be no customs officials. They drove through Obregón without stopping. Twenty miles south they came to a valley of wheats—brown, tangled, and rotting from stem rust. Borlaug stopped the truck and climbed down for closer inspection. They were the usual Mexican mixture of native wheats, devastated by *Puccinia graminas tritici,* the most prevalent of the major stem rusts —dead wheat as far as the eye could see!

As Borlaug stood there, the owner of the field came over the hill on an old, tired horse. "What are you doing on my land?" he demanded, his eyes flashing with anger.

Borlaug fumbled for his Spanish, then looked to Enciso, who climbed down from the cabin and came to his aid. There was a torrent of Spanish as he explained that Borlaug was from a great American foundation and they had just harvested a new crop of wheat in the fields north of Obregón, and it was being taken to Toluca for multiplication.

The farmer looked suspicious, and Enciso went on to explain that they had developed a new rust-resistant wheat that they hoped to multiply until they had enough for all the farms in the country. The man's eyes came alive as the words sank in, and he grabbed Borlaug by the arm. "You must give me some, please!"

Borlaug hesitated. Every seed was needed if he was to get the wheat into commercial production on schedule. But the man was insistent. "I have lost so much. I promise I will share the seed with my neighbors and family."

He wrenched his arm free and put several handfuls of the new varieties into a paper bag. The Mexican took off his sombrero and bent his head. Then in a sudden gesture, he gripped Borlaug's hand and kissed it.

It renewed Borlaug's belief in the underdog. Back in the heart of the Yaqui he would have to coax and cajole the wealthier men to plant test strips. Here this small farmer kissed his hand for his help. "He will probably keep all the seed at first," Borlaug told Enciso. "But he will brag to his friends and relatives and show them the new wheat. Then they will all want seed. If he were a commercial man he would see the potential in what I gave him. He could multiply it and make a fortune."

There was some prophecy in his words. In the years ahead there would be instances of newly developed seed taken from the wheat project without authority and sold without permission. There would be men trained in the program who would quit government work and take the seed and the know-how into commercial practice. One of these would be Teodoro Enciso himself.

Eight

High in the blue sky over Chapingo birds circled in a flock, then broke from tight flight into scattered groups, turning, swooping, rising again, waiting, waiting. They were hungry, but they were patient. Borlaug stood with Joe Rupert by the adobe hut and watched them hovering in the rare, thin air 7,500 feet higher than the sea.

Along the fresh black furrows Borlaug and Rupert had set the new generations of Supremo, Frontera, Kenya, and the newer Mentana and Marroqui crosses—the products of the Calles station in the Yaqui. The seeds would be sweet pickings to the birds, sweeter still when they had germinated into seedlings. In this crucial summer of 1946 Borlaug's new wheat crop was in danger. The danger was not so acute in the evenings, for Borlaug and Rupert were there until dark. But in the mornings, before the sun crept up over the eastern Sierra, the birds would be down among the rows—crows, grackles, jackdaws, and sparrows, scratching, eating, destroying.

"We need watchdogs," Joe Rupert said. "Human watchdogs. Suppose we get some old men from a village to come out here? Won't cost much and it would be worth it for the few weeks till the plants get set."

Borlaug thought it over. It wouldn't work. The fields

were too large; the birds would be too quick for old men. "I think we need boys, Joe," he said. "Bird boys! Young kids who will not just take the pesos and go to sleep in the field or forget to come one morning."

The project foreman, Manuel, knew the boys they needed. They were from his own village of Boyeras. They could be there at dawn the next morning, he said, but they would have to be paid five pesos a week, in advance, or their parents would not allow them to come.

When Borlaug and Rupert drove up the lane to the adobe hut at seven the next morning, the boys were already in the fields. Dressed in ragged, collarless shirts and frayed cotton pants, their faces lost under straw sombreros, the boys were small and quick and darted here, then there, on bare feet as arrogant birds teasingly, testingly landed on distant parts of the field. Each boy carried a short length of sisal rope; when he spotted a bird he ran toward it, whipping his rope in the air, making a cracking sound, and shouting until the bird flew away. Borlaug felt a tenderness toward these human scarecrows. He grinned at Rupert. "It's the end of easy pickings for those damned birds, Joe."

He called to the boys, who came slowly, shy and hesitant. He gave them sandwiches and cookies, which Margaret had packed, and asked their names. Their spokesman stepped forward. He had a small, square face, square shoulders, and a straight back. "My name is Reyes Vega," he said. "I am ten years old. These two are my younger brothers. But I am in charge, señor."

Borlaug put his hand on Reyes' shoulder. "You and your brothers do a good job. You will be paid every week. Come every morning and every evening, O.K.?

"O.K., señor."

They went down the path, toward the village, looking back shyly over their shoulders. "Did you see that kid's eyes,

Norm?" Rupert asked Borlaug. Borlaug nodded; he had noticed Reyes Vega's eyes—alive, alert, aware. The scientist could picture the boy's home. Mud walls and windowless little rooms; earthen floor and a straw mat for sleeping, perhaps the three brothers together; only the fire in the kitchen for heating, a pool for bathing, and a diet of tortilla and soup.

"Yes, Joe," Borlaug said, "I saw his eyes. Sharp and bright. He's a good kid. He'll do well, that one."

Now that the Chapingo wheats were planted and safe, Borlaug left Rupert in charge and drove the old Pontiac to Toluca. He took a small sack of the Supremo and Frontera seeds with him and about two pounds of the Kenya types. Richard Spurlock was away, but Mrs. Spurlock was in the kitchen of the Rancho El Cerrillo. She remembered him and asked about Mr. Rupert.

Borlaug wasted no time getting down to business. "Mrs. Spurlock," he said, tapping the bags of seeds he carried, "this seed is precious. It's for Dick and he knows it's coming. Have you got a really safe place to keep it?"

She looked a little puzzled. Then she said that the safest place was the iron strongbox built into the wall in her husband's office. That was really secure.

"But, Dr. Borlaug, I'm afraid it's full of money. Dick has to pay the workers tomorrow and there are thousands of pesos in there, notes and coin. There's no room for the bags."

Borlaug was almost bruque with her. "Take the money out, please!" he said. "This is far more valuable than a few thousand pesos. You can put the cash in a drawer and lock it up. This seed must go in that safe."

When Richard Spurlock returned home that night she told him how Borlaug had cleaned the cash from the safe and put the seed in its place. "What an astonishing man!" she said.

Spurlock nodded. "Yes. He is a bit unusual."

The seed was planted in the valley near Toluca. Spurlock had it all in the ground within the next two days. Then he went down to Chapingo and, with Borlaug and more seed, drove out to the farms he managed near Irapuato and other places in the Bajío. Borlaug told him as they rode along: "Most of the seed you're putting in is Supremo, Dick. Give it a feed of nitrogen and some phosphorus and it will do well. Most important, it will keep out the worst of the stem rust. That's why I'm asking you to plant it in the summer. Later we'll put it in with the usual winter wheats."

Spurlock took him to the little Bajío town of San Juan del Río and called on a farmer he knew. Borlaug gave him several pounds of seed, and Spurlock told him to put it in the ground, with plenty of fertilizer. The Mexican looked frightened.

"That is *veneno,* señor, poison! It can make the plants grow, maybe, but it will poison me and my family."

Spurlock took back the seed and they drove farther along the road. There was another farmer he knew, a larger land owner named Carlos Rodríguez.

"He's got more sense," Spurlock said. They gave Carlos enough antirust seed to cover two acres, and Spurlock took a bag of fertilizer from the trunk of his pickup and showed him how to spread it. They left him standing by his field with the bag of seed in his hand, the fertilizer at his feet, wondering what was to happen to him. They were to see him again in a year, a different man with a new hope.

As they drove back to the city, Borlaug worried aloud to Spurlock. "Each time we plant a seed the task gets bigger. We can't do it all. We're never going to have enough hands, or enough time. We need a hell of a lot more trained people.

We need extension services, people working out among the Carlos Rodríguez's of Mexico, telling them what to do, how to do it."

Spurlock did a calculation for him. Two more planting seasons, winter and summer, and he could reckon on having 20,000 tons of antirust wheats for planting in Mexico. Borlaug said: "I think we're on the right track, Dick. But we might not have that much time before we get a big rust attack."

That same summer Borlaug visited some of the northern areas of central Mexico, in the state of Coahuila. This was a region where the ejido system was least effective, where the mountain storms carried away topsoils and the moist weather nurtured rusts. He borrowed the Office station wagon for the 1,000-mile round trip and took Joe Rupert along to help him plant strips of selected antirust wheats.

It took three days to reach Saltillo, the capital of Coahuila. Founded in the sixteenth century and nestled in the foothills of the eastern Sierras, it had once been a seat of Mexican power, controlling a vast sweep of country, including most of present-day Texas. Now Saltillo was a sleepy city. They went straight to the small school of agriculture and asked if there was someone with English and knowledge of the area who could help them explore the high valleys to the south for test sites for new breeds of wheat. In a few minutes a young man appeared and introduced himself as Ignacio Narváez. He was a final-year student, he said, and would be honored to help them. He had a clear gaze, a flashing smile, and a strong, confident handclasp.

The trio set out at seven the next morning. They would spend several days in the region, returning to the city in the evenings. Narváez knew at once how to proceed. There was a family about sixty miles to the south, he said, who would surely give them permission to plant wheat.

The strip of land the farmer allotted them was in a gully well beyond the road high above the farmer's adobe house. He gave them a team of oxen and they worked through the day, leveling, furrowing, and seeding. Borlaug noticed where floods of past summers had exposed bedrock in places, but it did not prepare him for what was to follow. Clouds massed over the mountains beyond them, and big raindrops began splattering onto the soil. The flood came suddenly, racing down the steep ravines and gullies. Narváez called on them to be quick. They tied the oxen to a tree on high ground, then drove the station wagon back down the way they had come, toward the farmer's adobe hut. A hundred yards down the torrent cut their path. There was no way out and the water was still rising. Borlaug took the vehicle back to the high ground. The light began fading over the Sierras; the rain beat on the roof of the station wagon. They had no food, no coffee, no means of warmth. They stretched out in the back of the wagon with the two sleeping bags thrown over them, shivering together, dozing on the steel decking, waiting for daylight.

The morning brought no relief. Borlaug swore heartily. The ground they had planted the day before was a river of mud. On the second day the rain finally stopped, and by midday the water between them and the road was low enough to ford. The farmer's house was surrounded by water, so Narváez guided them several miles down the roadway to a small hotel he knew. The proprietor heated water for them and they took turns washing. Borlaug went last. When he came downstairs, ravenous for a hot meal, he heard a voice and a piano. Narváez was at the keys, singing and playing, not a sign that he had been stranded for two nights and two days without food. Borlaug was impressed. There was a future for such a resilient young man in the Office program.

They worked through the week on various farms, planting, trudging in the mud, sweating in the sun. Ignacio Narváez, or "Nacho," as they had taken to calling him, was a good worker, and the three men developed a close camaraderie and mutual respect. On their last night in Coahuila Borlaug said to him: "Nacho, how would you like to join us—come down to Mexico City and work with us in the Office group?"

Narváez felt the heavens had opened for him—to work with such a man, to study under him. But he said simply: "Dr. Borlaug, I am very proud you have asked me to join you."

Borlaug made the arrangements, and later in the summer, after his graduation, Narváez joined him. It was to be a rich association for them both—and for Mexico.

When he got back to the Toluca Valley and the Bajío, Borlaug visited the farms where the first of his antirust wheats had been planted; they were lank and unhappy. Tall wheats, too easily blown by the wind, their shapes told him they were not well. "This is damned puzzling," he said to Joe Rupert. "Look at those plants—yellow, wobbly at the knees. Yet for the first six weeks they were beautiful, dark green and healthy. Something is wrong."

Rupert asked how much fertilizer had gone into the land. Borlaug said he had applied about forty pounds of nitrogen to the acre, equal to the highest levels then used in the United States on wheat.

"It should be enough," he said, worried. He had too little data to make a reasonable diagnosis as to why the plants were not responding. The wheats at the Chapingo station, on the other hand, were green and healthy that August. He could only guess that the poorer soils of the Bajío were so badly depleted of basic nutrients that normal application of chemical fertilizer—nitrogen, phosphorus potash—was exhausted by the plants in the first six or seven weeks of growth. There was another reason for the failure of the Bajío wheats, but he was not to discover it until later.

One evening after work Borlaug went home to the apartment in Lomas to find Margaret troubled. When he asked her what was disturbing her, she told him she was pregnant again. It should have been an occasion for happiness, but the wounds from Scotty's tragedy were still fresh. They kept from each other their concern and deep fear of what the future might hold for this baby, but it was to haunt them through the months ahead.

That same week the three-man Rockefeller Foundation Agricultural Advisory Committee—Dr. Richard Bradfield, Dr. Paul Mangelsdorf, and Dr. Elvin Stakman—met with the Office team in Mexico City. Dr. H. K. Hayes, the plant geneticist from Minnesota, had traveled down with Stakman. Both Stakman and Borlaug knew Dr. Hayes was somewhat dogmatic in his views and could at times be explosive at breaches of orthodox practice. His reaction to Borlaug's two-generations-a-year wheat-breeding program was consistent with his character.

When he saw the Yaqui Valley wheats growing well and sturdy at the Chapingo station, he was unimpressed. "What in hell, Borlaug? Didn't you learn the first lesson I taught you in your sophomore year? You're going to be going in circles with such a program—one step forward, the next step backward."

That evening, as Borlaug sat with Joe Rupert facing the committee in the Office rooms at San Jacinto, the plan met with more opposition. It was argued that Sonora was a long way from their base of operations; that it did not need as much assistance as the Bajío did. But Borlaug was adamant. "I tell you, what happens in the valley in Sonora is important to the entire wheat program; it will point the way for all Mexico. I want the authority to delegate a good share of our effort to that region. And I want to do it in the way I think is best."

Stakman could see the fire begin to flash in Borlaug's eyes, but before he could interject Harrar broke in. "Norman, I have told you each time the subject has come up that you should concentrate your efforts in the Bajío area. This scheme of yours has been considered twice, at your insistence, and voted down twice. Why can't you accept that?"

Harrar's abrupt dismissal sparked a fuse. Borlaug rose to his feet, slowly and deliberately. Stakman could see him fighting to contain his anger. He was hurt. "O.K., you have made your decision. You've decided where the hell the emphasis will be on this work—work you have charged *me* with carrying out! But you did not consult me. You did not consult Joe Rupert. You tell me, 'Go and grow more bread for Mexico.' But you tie my arms behind my back and say I must do it only here, not there. Well, that is not my way!"

He paused and leveled his gaze at them. "And now, if this is a firm decision, I also make a firm decision. You will have to find someone else to conform to your rules—"

Stakman broke in. "Now, Norman, don't go jumping the gun."

"I have a couple more things to say. You're laying down a policy that is wrong, and I can't go along with it. I'll stay until you get a replacement. If you want Joe Rupert to take over, that's O.K. with me too. But as of now I resign. You'll have it in writing first thing tomorrow."

He turned and walked away from the table. The room was tense and quiet. He was through the door when he heard a chair scrape back. Joe Rupert's voice came ringing out of the room like a bell: "And that goes for me too, damn it!"

Margaret Borlaug was disturbed by the turn of events. Her

husband and Joe had stormed into the apartment incensed and affronted by the way things had gone at the meeting. Borlaug told Margaret he had taken the only possible course. Harrar had shown lack of trust in his judgment, and he could not go against his own convictions just to hold his job. But Margaret believed they had acted in anger—and therefore, perhaps, in haste.

Borlaug still held the handful of mail he had snatched from his tray as he and Rupert stormed from the office. As Margaret argued with Joe, he looked at the envelopes, and one caught his eyes. It was handwritten, in a spidery scrawl, and had a Sonora postmark. He opened it and saw the signature of Aureliano Campoy. He could see again the old man's weathered face in the shadow of his sombrero.

It was a brief note, attached to a copy of a letter Campoy had written to Dr. Harrar. The note to Borlaug said he had sown the seed he had given him on two acres, and "the result has been better than anything I ever saw on my farm. It is wonderful wheat and I shall keep it for seed next year. From my heart I congratulate you."

In his letter to Harrar, Campoy expressed admiration for the Rockefeller Foundation and for its mission in Mexico:

"For the first time, Dr. Harrar, I have seen a man come here from the Rockefeller Foundation or any other organization to help the farmer. Perhaps it is the first time in the history of Mexico that any scientist tried to help our farmers—I don't know that. The results are already evident in my own land with this new, wonderful wheat. I thank you, and I thank Dr. Borlaug. But why is it, with such a great force like the Rockefeller Foundation, that you do not give your men the tools and machinery they must have to fight with? Why does he come like a beggar to borrow the tools to grow new wheat? You fight hunger, yes! Is it not time somebody was fighting to help you inside your own organization? I want to say what is happening here with

Dr. Borlaug will have a tremendous effect within a short time."

Borlaug was sure Mrs. Jones had helped him with the letter. But what would it accomplish?

"We'll see tomorrow," Borlaug said. "We'll know whether this means anything to Harrar."

Margaret looked at him. "But," she reminded him, "you've resigned! So what difference does it make?"

Borlaug and Rupert were at the San Jacinto premises at six-thirty next morning to write their resignations. Dr. Stakman sat by their desks, glowering at them from under his brows. Borlaug had never seen Stakman at that hour of the morning.

Stakman rebuked them for flying off the handle. "Hell, Borlaug! You and Rupert behaved like spoiled children last night!"

Borlaug wondered why he said nothing about the letter from Campoy. He knew that Harrar would have shown it to him. Stakman had in fact seen the letter, but it was not an issue with him. He was concerned with tactics.

"The first thing to get into your angry head, Norman, is that I am *not* in opposition to what you want to do! I am here to ask you: Are you really convinced you are right?"

Borlaug said: "Of course I'm right! Why else would I resign?"

Stakman waved a placating hand. "If you think you are right, why in heaven don't you see that you get your way? Don't you see? You should stay here and keep fighting until you get what you want! Don't walk out and leave the field to your opposition. Sure Harrar talks tough. He's a strong man, and he can't give way easily. But he gives you a lot of freedom of action and initiative, and until you have been absolutely, specifically prohibited from doing what you feel must be done in Sonora—there are always ways of getting around things like this. Your enemy is not here, in this office. Your enemy is rust—out there in the field. I know Hayes and others have got you riled, but that's no reason to behave like a petulant child."

Borlaug was still angry. "They are making an impossible task, to define such limitations. They can tell me what they want done; they can't tell me how to do it," he argued.

He went to his desk and took a sheet of paper and a pen to write the resignation. Stakman came and stopped him, putting a hand on his shoulder. He still had influence over his former student.

"Do this for me, Norman. Don't write that until tonight. Give me a chance to talk to Harrar, and let a day of work go under your belt."

Borlaug could not refuse.

When they came back to San Jacinto from Chapingo that night, Stakman and Harrar were waiting in the office. Stakman was smiling, holding an unlit cigar in his fingers. Harrar got to his feet and shook hands with Borlaug. The matter had been reconsidered, he said, and the way was cleared for Borlaug to devote half of his efforts to the north. Resources, equipment, money, and men would be made available. It would be a recognized sector of the Office policy and part of the wheat project. He now hoped that he and Rupert would see fit to withdraw their verbal resignations, forget the clash of opinion, and get on with the mission. The crisis was over. Stakman had made the peace. It was fortunate he had been there. Borlaug was always certain he would not have given way in the rough and tumble. He would have gone through and quit the job.

They went to Stakman's hotel and ordered tequilas. As they talked away the tension, Stakman said: "I guess we all behave like boys at times."

Margaret was pleased when Borlaug told her how the differences had been settled. "It's good, Norman. You would have been unhappy to leave here."

She knew that it would mean longer hours, longer times away from home, but she understood.

Harrar's approval for Borlaug to expand the wheat program in the Yaqui Valley cleared the way for the next moves in his strategy against rust. In the fall he freighted two bags of the newest seed for winter planting and flew to Ciudad to Obregón. He lost no time calling on his friends. He thanked Campoy for his letter. He believed it had influenced Harrar. He gave him seed, and did the same with Mrs. Jones and Roberto Maurer, who undertook to pass some along to other farmers for planting in the Yaqui soil. It would all be harvested in the spring and be gathered in as seed, not as food grain. The various farmers would then keep some for their own planting, and when there was enough it would be fed into the government's newly organized seed agency for distribution.

He was back in Mexico City before Christmas and in the following weeks made journeys into the central areas with Rupert and several of the new Mexican trainees. The weeks went by as he explored the country further and devoted himself to training and teaching. The winter crops were sown and doing well, with smatterings of the Supremo, Frontera, and Kenya types in small lots.

Late in March Margaret went into the maternity home in Lomas for the birth of their third child. It was a boy, born on March 29, 1947. He was healthy and their unspoken fears were dispelled. It was beautiful for Norman Borlaug to see Margaret's face as she held their son in her arms. They decided to call him Billy. His coming filled out their lives and rounded off the start of what seemed to be a good year.

The year 1947 was not without its temporary setbacks, however. First, the fall harvest was threatened by rain. Big black clouds rolled in from the Pacific, bringing storms that turned the wheat fields of the Bajío into quagmires. In the Toluca valley Richard Spurlock's forty acres of Supremo and Frontera wheats were ready to harvest, but he had not a hope of taking his machinery into the fields; the vehicles would have sunk to the axles. There was no sign of letup, and Borlaug knew he could not wait or he would lose his largest crop of commercial seed grain to date.

He got Joe Rupert and they went out with hand sickles. Every morning they worked with large crews of laborers on the wet harvest, slashing the wheat, binding the sheaves, and stacking them to dry in Spurlock's big thatched barn at the end of the field. Every afternoon the rains would soak everything again. Spurlock put a dozen peons to work turning the sheaves to keep them from rotting or mildewing. Finally, after fifteen days of backbreaking work, all the grain was in. They threshed the wheat and bagged it; the Spurlock holding had produced fifty bushels an acre—at least four times the average yield that year. "You know what this means, Dick?" Borlaug said. "We're on our way! We'll have that 20,000 tons of seed for the farmers by next year."

He left half the seed with Spurlock to plant as a winter crop and took the other half to divide between the Chapingo and Yaqui stations. The Chapingo planting was uneventful, but the trip north to Sonora was a near disaster.

Now that he had Harrar's official approval, Borlaug was able to get two pickup trucks for the journey and a full crew: Joe Rupert, Teodoro Enciso, and two new trainees, Alfredo Campos, a Mexican graduate, and Oscar Nery Sosa, a Guatemalan exchange student. Borlaug would drive the lead truck, Rupert the second. They planned to take the coast route—across from Durango, on to the new road being cut over Devil's Backbone in the Sierras, down to Mazatlán, and on up the coast to Sonora. Early in November they set out, one truck carrying a small tractor and equipment, the other the seed and enough food for the trip. They had a clear run to Durango; then the weather turned bad and the roads deteriorated steadily as they climbed higher into the mountains. On the morning of the second day it began to rain, and the unfinished road surface turned into a greasy cover of wet clay. They were on a narrow stretch of dirt, cautiously rounding a bend of Devil's Backbone, when Borlaug braked to a halt. Ahead their passage was blocked by a truck, and beyond that the road disappeared into a sliding mass of clay. There was a sheer drop on one side, a steep slope on the other, and the road was too narrow to turn around. Borlaug and Enciso went forward to talk to the driver in the truck ahead. It had been there for two hours. It belonged to the Mexican highway authority, and the driver said that when he and his partner failed to arrive at the depot about 150 miles farther on, highway officials would most likely send a bulldozer crew to look for them. Borlaug walked back along the trail and around the bend to see if there was any way out of the impasse; suddenly he heard a rumble and felt a trembling. Then he saw the landslide slipping down the surface of the mountain. In a moment the road behind them was also blocked.

They spent two days on the dank face of the mountain before they heard the clank and rumble of a bulldozer. The heavy machine slowly pushed the tons of clay down the slope, and they slid away with a sucking, squelching sound into the deep canyon. Borlaug looked down and shivered.

The three trucks made their way precariously over the wet surface cleared by the bulldozer. They still faced a long, miserable descent through forbidding country, and the fine rain had settled into a constant drizzle. Five hours later they reached the highway authority depot and asked about the road ahead. The officials had a glum story. Heavy rains in the Sierra had caused exceptionally heavy flooding. There were bad washouts all along the route, and it would be weeks before the road could be opened. There was, however, one railway line from Mazatlán to Nogales in the north that was still open—for how long they could not say.

Borlaug made his decision on the spot. They drove to Mazatlán, where Teodoro Enciso and Alfredo Campos packed up the seed and boarded the train for Nogales in Sonora. From there they would hire a truck to the Yaqui station, borrow some equipment from Campoy, and get the seed into the ground immediately. Borlaug, Rupert, and the young Guatemalan would drive the pickup trucks back to Mexico City, and Borlaug would then fly to Ciudad Obregón to rejoin the Mexican trainees.

They decided the trip back to Durango via the Devil's Backbone was impossible, so they headed south from Mazatlán on the coast road toward Guadalajara. But as they started south the roads were a sea of mud. Finally they came to a river where two women sat washing clothes, and Borlaug asked if any vehicles had forded it that day. One of the women said a truck had gone through, and Borlaug proceeded, with the Guatemalan student, in the lead truck. The river was about 150 yards wide, and halfway across there was a surge of water. It rose silently, creeping over the running board and then swirling around their ankles. The engine stopped. The Guatemalan boy said he could not swim, and Borlaug told him to stay in the truck. He took off his leather boots and climbed down into the water. It came up to his armpits and was flowing faster now. On the far bank he could see what looked like a road-building camp and he made for it. The river bed was covered with sharp stones that cut his feet as he tried to keep his balance against the strong current. Suddenly he tripped and was swept downstream in the torrent.

Somehow he managed to swim ashore; he clambered up the bank a few hundred yards downstream and stumbled toward the road-building camp. Borlaug told the workers that there was a government truck in the water and that a man's life was in danger. He needed their bulldozer for a rescue. The foreman said the engineer was the only man who could give them permission to use the machine, and he was in the next village. Borlaug had to plead, then threaten, until the foreman agreed to take a cable and run the bulldozer down to the river. By that time the water had risen above the seat of the cabin, and Oscar Nery Sosa was sitting on the roof, calling to them for help.

It took nearly a dozen men to hook the cable onto the front bumper, and finally the bulldozer started to pull the vehicle. As the wheels were jolted free, the force of the river began to tip the truck. Borlaug called out to the men in the water. They threw their weight to the side of the truck to hold it against the current as it was slowly hauled to the safety of the bank. It was late at night, and they had to wait until morning to haul the other pickup across. But even then they could not go on. The waters had ruined the oil in the engine of the first truck and it was immobilized. The Mexican engineer put two of his men to work greasing it and cleaning off the slime, and finally they were on their way.

By the time they reached Mexico City they were exhausted. When Borlaug and Rupert arrived at the apartment in Lomas, Margaret was shocked at their condition. They bathed and went to bed and slept for sixteen hours. When they awoke she called a doctor, who dressed Borlaug's cut and swollen feet and ordered him to rest for a week.

Borlaug shook his head. "Impossible," he said. "I've got to get up to see what's going on at the Yaqui station. Our whole work swings on that seed and what it produces next spring."

Margaret knew that argument was fruitless. She could

talk for days without changing his mind. Next day Borlaug booked a flight to Ciudad Obregón and was soon on the road toward the Yaqui station. It was with immeasurable relief that he looked over his experimental fields and saw the neatly furrowed rows of newly planted wheat.

Nine

For a long time Norman Borlaug had wanted to introduce the farmers of the Yaqui Valley to the advantages of modern farming technology. He was particularly anxious to have those with major influence, such as Don Rodolfo Calles, view the fruits of his labors. Once they saw first-hand the superior wheat that could be grown through modern research and the application of modern cultivation techniques, he reasoned, they would give him their full support. He had hesitated simply because it took time for his labors to bear fruit. But in the spring of 1948 he decided that he was ready.

He called on Roberto and Teresa Maurer, and together they planned their strategy. Teresa sat down and hand-lettered two dozen invitations in her neat Spanish script. In addition, Roberto knew a government official in town, Inginero E. Reza Rivera, who might help spread the word, and he would try to persuade a few of his own friends to come.

Borlaug borrowed a flat-top truck from Campoy and trestles and tabletops from Mrs. Jones. On the big day he was ready with a program of tours of the experimental plots and explanations for the visiting farmers. He sat on the back of the truck with his assistants, Rupert, Campos, and Enciso, waiting for the first of the farmers.

By midday there had been only five visitors: Aureliano Campoy, Mrs. Jones, Roberto Maurer, and two of the Maurers' neighbors. Borlaug waited all afternoon for some of the big farmers—Don Rodolfo, Don Jorge Parada, Don Eduardo Vargas—but they did not come.

He considered it an affront, particularly in the case of Don Rodolfo, and he was deeply resentful. He reasoned, however, that it would be woefully wrong to storm into Calles' home and demand to know why he had not attended the field day. He had to wait and then make his move.

He allowed a week to elapse, during which he traveled the length of the valley to see how the commercial wheat crop was developing and how the land was being extended by the expansion of irrigation. He noticed that in many fields of the native criollo mixtures rust was gaining a hold. He decided to use this as a lever with Don Rodolfo.

He drove to the Calles farm and was received as before. There was the same air of polite surprise, the reserved manner. Borlaug said he had taken advantage of his invitation to call again and opened his skirmish by telling Don Rodolfo about the signs of a returning rust epidemic. He pointed out that many farmers were already losing their profit—some at least a quarter of their harvest. There were pockets of infection along the valley, and under the right conditions they could spread. This meant that Don Rodolfo had a community obligation that he was not accepting.

The black eyebrows were raised in polite inquiry. Borlaug went on with his blunt challenge. They were working on new wheats at the research station that would protect the farmers against the menace of rust. It was time Don Rodolfo and the other leading farmers took an interest in this work and fostered its application.

Don Rodolfo was looking at him with new interest, and

Borlaug thought for a moment he had won a convert. But the former governor got to his feet. "Come with me, Dr. Borlaug. You think only you have new wheat? I will show you a field of *real* wheat. It will give a bigger harvest than anything you have up there."

He had traveled to the United States and had seen wheats with big golden heads. Now they were in his fields, nodding and tall, graceful and full of promise. Borlaug recognized the type at once: a Texan wheat known as Austin. Some of its line had been crossed with Supremo, and it had a resistance to certain rusts. But it had a fatal weakness. "I know this wheat," he told Don Rodolfo. "Be careful. It has a weak neck. When it is heavy in the head, near to harvest, and the winds come, the necks will break and your seed will be on the ground."

Calles laughed pleasantly, as if to say he knew pique when he saw it. He patted Borlaug's shoulder and bade him farewell.

A week later Borlaug and Enciso were driving to the airport to ship their bags of seed back to Mexico City. Down the side road, in the distance, Borlaug could see activity in the Calles wheat fields and turned off at the property.

There were two foremen and a gang of workers. Don Rodolfo was shouting to them: "Hurry, hurry! We have no time to lose."

Borlaug walked across the field. "What goes on here, Don Rodolfo?"

Don Rodolfo looked at him, flushed and angry. "Yesterday these wheats were beautiful. This morning they are falling in the wind. See them? Look there—and there. Hurry, men, hurry!"

The heads were falling quickly, snapping off with too much weight and the weakness of the straw neck. Food for the mice and the insects. The machines were rolling now.

Calles' concerned voice stopped booming across the fields; he dropped a hand on Borlaug's shoulder. "I shall remember when you tell me things in the future, my good doctor. You shall see."

Calles kept his word. The following spring, when Borlaug staged his second field day, Don Rodolfo appeared with a group of Yaqui farmers, among whom Borlaug recognized Don Jorge Parada and Don Eduardo Vargas. They were serious and attentive as Borlaug explained the potential impact of his high-yielding rust-resistant wheats on Mexico's future grain production. When he first came to Mexico, he said, wheat production was only half of what was needed, about 330,000 metric tons. With the new wheats he believed it would rise to 575,000 metric tons in the next year. In another four or five years Mexico could be self-sufficient—provided the rust was held at bay and the soils were fertilized properly.

After the speech Don Rodolfo came up to shake his hand. "You spoke very well, doctor. It looks to me like this event will become an annual affair in our Yaqui Valley. We ought to make it more of a social event. Perhaps next year we can organize a small committee to bring chairs and tables and refreshment of some kind, eh?"

Borlaug had won a major success in the Yaqui Valley. But meanwhile an outbreak of stem rust was spreading through the Bajío country. Borlaug went out to the stricken areas, taking Joe Rupert and Richard Spurlock with him. The rust had cut through the native criollos like a scythe. The fields smelled like sour milk, and there were wide panoramas of ruined crops.

But not all the wheats succumbed. It was a baptism by fire for Borlaug's resistant varieties. In some areas of the Bajío he saw them standing, isolated among the matted wreckage. They were fresh and untouched, ripening gold and green, defiant of the floating pathogens of rust in the air.

The three men stopped near the farm where two seasons earlier Borlaug had given a bag of seed to the farmer Carlos Rodríguez. As they stood looking at the Supremo wheats, Rodríguez appeared from behind his house. Calling loudly to his family and neighbors, he ran down to the road and fell on Borlaug, embracing him with both arms in Mexican greeting, his black eyes unashamedly running with tears. He was babbling in Spanish, too fast for Borlaug to understand.

Other men came running. They too embraced him. Then the women followed, more hesitant, restrained.

"What's this all about, Dick?" asked Borlaug. "They don't even know me, these people!"

Spurlock explained the reason for the excitement. Carlos Rodríguez had taken the seed and the fertilizer they had given him and had grown three times more grain than he had ever gotten from his little farm. He did what he was told. He did not eat the grain but planted it again at once. He used more fertilizer, and from that crop he gave seed to his neighbors. When the fearful rust disease struck, their crops were the only ones not touched. They regarded it as a gift from heaven, and Borlaug was the agent who brought them that gift. They believed their lives would be changed now. They had hope and their reaction was understandable.

"They are just saying 'thank you' in the best way they know," said Spurlock. "It is a simple thing—a matter of having something to eat or going to bed hungry."

That same summer the first Latin American science conference was held in Mexico City. There were agricultural scientists from Brazil, Uruguay, Argentina, Chile, Peru, Bolivia, and Colombia, and they saw some of the Rockefeller program at work. During the conference Borlaug talked about the success of the wheat development program, citing the recent devastation of the native wheats by stem rust and the resistance of the Hope derivatives and Kenya varieties.

"This land of Mexico can now move on to produce much more wheat," he told them. "We have a shield against common forms of rust. Our next step is to introduce Mexican farmers to the use of fertilizer. Much greater returns will be won from this country if the poor soils are fed. And that probably also applies to many countries of South America."

The words triggered reaction, for within weeks Harrar received an official request from the Colombian government asking if the Rockefeller Foundation would send two of their senior scientists to look at the problems in that country. Foundation headquarters in New York approved, and Harrar asked Ed Wellhausen and Borlaug to make the journey down to Colombia. They spent three weeks studying Colombian agriculture and visiting government stations, experimental institutes, and colleges. It was the same pattern as that in Mexico: primitive farming—and poverty among the people.

But an essential difference in the land made the backward nature of Colombia's agronomy more culpable. There was no shortage of moisture: unlike Mexico, Colombia did not face the problem of arid and semiarid soils. But this wetter climate would aid rust and blights, so if the diseases were controlled, nothing but lack of action would stand in the way of increased food production.

They said all this in their report to the Rockefeller Foundation, which they submitted to Harrar on their return. They suggested that Colombia could gain greatly from scientific aid and recommended the introduction of a program similar to that in Mexico. The New York headquarters studied the report and suggested that Borlaug might be interested in leading the Colombian project. But he could not leave Mexico just when his wheat program was showing results, so the foundation appointed Dr. Lewis Roberts, a corn breeder, to take on the assignment. Joe Rupert went with him as wheat breeder and pathologist. There were arrangements to be made and facilities to be organized, but by early 1950 the Colombian program was under way. It was the Rockefeller Foundation's first venture into South American agriculture.

The following year brought a major setback to Borlaug's project. A new strain of rust swept through Mexico. Named 15B, for its deadly effect on fifteen widely grown wheat varieties, it had appeared first in the southern United States in the summer of 1950, destroying millions of acres of wheat. Now it spread to Mexico, hitting Borlaug's Supremo, Frontera, and related types.

It was a shock to government plant pathologists, both in Mexico and in the United States. In Washington Dr. H. A. Rodenheiser reacted by organizing the first international stem rust nursery. With the cooperation of Borlaug and others, he initiated a network of testing centers in many countries, in an attempt to provide the world with early data on the emergence of new strains of rust that could menace man's food supply.

That same year, 1951, Borlaug was asked to join Dr. B. B. Bayles and other agronomists with the U.S. Department of Agriculture in a survey of South America. For six weeks they toured various countries: Chile, Peru, the great expanse of Brazil with its tremendous soil problems, the vast devastation in the grain belt of central Argentina (which had been stricken by stem rust), Paraguay, Bolivia, Ecuador, and Colombia. Everywhere there was the problem of antediluvian agriculture coupled with plant diseases of great variation.

He was close to Bayles throughout the trip, and they talked for hours, sorting out facts and impressions, seeking answers to the disease situation. Bayles spoke of the complexity of the pattern. The diseases were not universally prevalent but dominant in different areas. There was a clue somewhere in that fact. Plants of the same breed could be overwhelmed by rust spores, he mused, while similar wheats would resist.

His words rekindled an idea that Borlaug had first considered during the rust attacks in Mexico earlier that year. Now he tested it on Bayles. "Suppose I tried to breed a composite wheat, selecting strains with similar habits, type of spike, same size and color of grain, same growth patterns, same height—but with a wide variety of disease resistance?"

Bayles thought he would not get very far. Nature would rebel, he said. The best traits in the wheats might easily be lost in the rearrangements of the germ plasm.

What if he played a game of numbers, crossing and recrossing the best varieties on a vast scale, always being extremely selective?

"It would be a mountain of work," said Bayles, "but you might just get away with it." It was enough encouragement to renew Borlaug's enthusiasm for the idea.

They discussed other problems. There was the difficulty of persuading the farmer to enrich his soils with chemical fertilizers and, beyond that, the problem of the ceiling on yields. At some point fertilization became counterproductive: the wheat stalks would grow too tall and slender, and the wheat would lodge, or topple, of its own weight before it could be harvested. What was needed was wheat with a stronger straw.

Bayles mentioned that Dr. Orville Vogel, a plant breeder for the U.S. Department of Agriculture, working at Washington State University, was crossing a new line of wheats from Japan—wheats with a short, strong straw—but they had a lot of drawbacks. It was the first Borlaug had heard of the Japanese "dwarf" wheats, which were to play such a major role in the Green Revolution. For the moment he was caught up in the idea of multilineal breeding against rust attack.

When he returned to Mexico City, Borlaug found many changes at the Office. Harrar had moved to Rockefeller Foundation headquarters in New York to direct new international projects, and Ed Wellhausen was now in charge of the Mexican operation. New teams were working in soils, insects, and beans, as well as in corn and potatoes. Dr. John Niederhauser was managing the potato project and was living near the Borlaugs in Lomas. Borlaug's own group had been expanded, and among the agrónomos now with him was the young student he had found in Saltillo, Ignacio Narváez, who was working toward his doctorate.

They had now trained more than 200 Mexican agrónomos, and many had gone on to higher education in the United States under the sponsorship of the Rockefeller Foundation. Some were already returning to their homeland, infusing fresh blood into agricultural teaching and research. A new college of agriculture had opened in the Sonora capital of Hermosillo, bringing the national total to four, and others were in the planning stages.

There was also a change in the premises of the Office. The old building in San Jacinto had become a pawn in a struggle between militant students, who wanted it as a dormitory, and the ministry of agriculture. The students threatened to burn the building unless it was vacated, and Wellhausen looked for a new headquarters. He finally found suitable space in Londres in the heart of the city, and the project was moved there.

Now Borlaug was able to spend some time with Margaret and the children. Jeanie was growing fast; blonde and talkative, she spoke Spanish like a native. There was some worry about Billy, who was subject to allergies and occasional asthmatic bouts. Borlaug was particularly concerned. All the boy would have to do was cry once in the night and he would be instantly awake and out of bed running to his side. He was solicitous about Billy's health for many years —such as he had never been with Jeanie.

He and Margaret had never had a worry about their lovely daughter; she grew straight and tall, vigorous, and self-assured. From the days when she had been able to prattle Jeanie had dealt with life's problems with aplomb. Borlaug thought that much of Margaret's practical approach had rubbed off on her. She attended a mixed school for Mexican and American children, and Billy had the benefit of her natural ability in language. She would often play the role of teacher with Billy, aping the mannerisms of her school instructors, wagging a finger at her pupil until he could sound the Spanish vowels correctly. These days laid the foundation for her later career as an outstanding Spanish teacher in American high schools.

There was a protective, mothering quality about her as well as a hesitant adoration of her father. Sometimes she would come into the room where they sat to talk about some argument with Billy and, while relating the situation to her mother, would look at Borlaug out of the corners of her eyes.

But because of his long periods away from home Borlaug often felt like a stranger to his children. He was always patient in trying to break through their shyness with him, playing ball with Billy for hours and talking to Jeanie in Spanish.

When Billy's birthday came in March 1952, Borlaug reflected that it was nine years since Harrar had opened the Mexican program and seven years since he had started his work there. It had been a long haul, but the Mexican wheat program was about to bear fruit.

That season Borlaug extended the number of distinct and separate wheat lines in his nurseries to more than 40,000. The log of work included a staggering 6,000 individual crossings of wheat heads.

Many plant breeders found the achievement hard to believe, but it was on record. What was not on record was the backbreaking effort, the discipline and patience needed to achieve these results. The concept of multilineal defense against rust in a single composite variety of high-yield wheat drove him on. He was compelled to enlist every pair of hands he could to work in the fields in the Yaqui Valley, including those of the fourteen-year-old former bird boy, Reyes Vega. In addition to the demanding physical effort of crossing the chosen 6,000 heads, every line had to be recorded in the swelling pedigree books. It was monumental labor, but it was worth it to Borlaug, for it offered the possibility of a major victory against rust, against hunger.

The memory of the swathe of destruction caused by the 15B rust was vivid in Borlaug's mind. Of his four initial rust-resistant wheats, only two, Kenya Rojo and Kenya Blanco, had withstood the attack. There were new, untested varieties—Kentana 48, Yaqui 50, Mexe, Bajío, the Chapingo lines. But the Kenya strains were the only rust-resistant wheats in commercial use.

Suddenly as rust again changed its face, some of these, too, succumbed. The enemy was classified as race 139; it had been known for twenty years as a mild and unimportant strain of stem rust that habitated barberry bushes in the United States. Then it changed its nature slightly and became a menace.

It was a shattering attack and made Borlaug's work on multilineal defense all the more urgent. He found the Kentana strain and the new Yaqui variety to have a combined defense against 139, and he crossed and recrossed them to strengthen their resistance. It seemed an endless circle of work.

Side by side with his fight against rust Borlaug was conducting a campaign to enrich Mexico's soils. More than 90 percent of the country's land was seriously depleted of essential nitrogen and other natural elements. Yet as late as 1950 the use of fertilizer was virtually nonexistent. It was essential that this situation be changed quickly. For one thing, Mexico's population explosion was steadily widening the gap between production and consumption. For another, wheat and corn might soon be imported cheaper than it could be grown on Mexico's own depleted lands. The country would then face a bleak future, with a stringently reduced agriculture and rampant unemployment.

Borlaug received strong support from Wellhausen in the struggle. Together they solved the mystery of why high-yield wheats and corns, fed at U.S. fertilizer levels in the Bajío, had not responded as they should. They engaged a tractor driver to spread their fertilizer on a test farm in the Bajío. When the wheat was heading, Borlaug called Wellhausen to this special field.

"Just have a look at that, Ed! What do you make of it? We put forty pounds per acre of nitrogen and phosphorus fertilizer on this field. Now look!"

Squares of lush, strong growth dotted the otherwise pallid field like a chessboard. Wellhausen was intrigued, then suspicious. He went to find the man who had put down the fertilizer. "Just tell us, move by move," he said, "exactly how you spread the fertilizer for this wheat."

It took some time to get to the truth. The man had not

followed Wellhausen's orders to spread the fertilizer evenly, working from the center outward. Instead he had gone up and down, and across and back, leaving squares where the fertilizer was two or more times the amount deposited elsewhere. That was the key: in the Bajío wheat needed at least twice as much fertilizer as in America—perhaps three or four times as much.

"Damn it, Norman," said Wellhausen. "Now we can show these disbelievers they can double their harvests. If we *told* them 20 or 30 percent increases were within their grasp, we'd have trouble convincing them. But *show* them, and they'll be running at the mouth."

Demonstration became a powerful persuader. The little men, the Carlos Rodríguezs of the Bajío, would all prod one another. Envy was a motivation. But in the Yaqui the big, influential men along the canals grew wheat with double or more the yield of the rest of the country. What would their reaction be to this soil enrichment argument?

In the spring of 1953, with Campos and Narváez and several other Mexican trainees, Borlaug staged his most successful field day in the Yaqui. He showed the farmers around the station and brought them to his finest display: tall, heavy-headed, healthy wheats of the new broadly resistant Kentana, Yaqui, and Chapingo varieties, into which he had crossbred the rust defenses combined from the Hope and Kenya types. He spoke in Spanish:

"Friends, this stand of high-yield wheats is the finest yet grown on this station. It is the most broadly resistant to disease. It will give you forty to fifty bushels of grain to the acre—if you feed it, as it has been fed here, with fifty pounds of fertilizer an acre."

There was silence; the air was pregnant with disbelief. He waited for questions that would challenge his assertions. It was Don Jorge Parada who first took issue.

"Dr. Borlaug. The seed you have given us from your

work is growing wonderfully well in this great valley. We are all in debt to you. We are grateful to you for what you have done for us, but you really must try to understand the situation here. Our soils do not need such help. This is new land, not tired land. Why should we throw our money away?"

They all nodded in agreement. An idea that cost money was hard to promote.

Borlaug countered: "You will not throw money away by putting fertilizer on your land; you will be throwing money away if you don't put fertilizer on your land! If there were gold in your land, you would dig it out! There is gold there. Put the fertilizer in and the gold comes out in the heads of your new wheat."

They were unmoved. Finally he told them to test it for themselves, to plant a small piece of land with the recommended fertilizer and compare the yield with their regular crops. It was a direct challenge and they could not resist. They agreed to give it a try next year. It was all Borlaug needed. The results, he knew, would speak for themselves.

There were to be some unfortunate repercussions, however, from the farmers' acceptance of fertilizer. When Borlaug returned to the Yaqui the following fall at sowing time, he was greeted with the news that a little-known American fertilizer company had set up headquarters in the farmers' cooperative in Ciudad Obregón. Curious as to the company's motives, he went to Roberto Maurer, who was president of the cooperative, and asked him what the story was.

The senior executive of a U.S. East Coast company had called on Roberto with a plan to bring a group of soil scientists, analysts, chemists, and plant pathologists to the Yaqui to study their soil needs. If Roberto would allot them room at the farmers' cooperative building, they would set up a small laboratory and do on-the-spot examinations of the soil of each farm so that every man would have his own formula and not have to waste money on fertilizer not suited to his particular needs.

Roberto had consulted officers of the cooperative and they agreed. Soon keen-faced young men had moved into the rooms, set up glass tubing, retorts, and vials, donned sparkling white coats, and started work. As each farmer called, they drove out to take samples of his soil to put through a complex series of tests. A day or two later the farmer was handed a formula calculated to fit the precise needs of his soil and was invited to order the formula through the company. He was urged to act swiftly, since delay in ordering might mean not having fertilizer in time for the winter wheat season.

The approach was effective. Roberto himself had asked for analysis of his land and had placed an order, cash in advance, for 150 tons. How much had he paid? Borlaug asked.

"A little more than $1,000, American," Roberto told him.

Borlaug frowned. "That's darned expensive, Roberto!"

Roberto argued that it would actually be cheaper in the long run, because he had his own formula now and would know exactly what to order in the future. Borlaug asked for the formula. Roberto brought it to him and his fears were confirmed. It was cheap, low-grade fertilizer with only 2 percent nitrogen, and nitrogen was the major need of the Yaqui soils. Applied by the tens of tons in the Yaqui the fertilizer would have minimal value.

He told Roberto grimly: "This fertilizer is useless here! This is a hoax, a selling trick. But much worse, it will destroy the trust and confidence we have built among thousands of farmers. Imagine a poor ejido farmer putting his family's money into this rubbish after I persuaded him to use fertilizer. He would turn his back on me forever!"

He asked who had ordered. Roberto knew of no important farmer who had not. All the big men along the valley had come in, including Don Rodolfo, Parada, Vargas, the Obregóns, and many others. Borlaug sat down in Roberto's cane chair, his shoulders slumped. "The fat is in the fire, Roberto."

Roberto telephoned the cooperative offices and spoke to a local Mexican girl who was acting as secretary. No—the chief executive was not in. He was out in the countryside but would be there tomorrow, and she would make an appointment for Roberto and Dr. Borlaug to see him at nine in the morning. It was a miserable evening for Borlaug.

Next morning the two men waited for the chief executive of the fertilizer company to appear. The secretary came, opened the doors to the rooms—and found a note awaiting her on the desk. The glassware, the fittings, the young men in the white coats had vanished. The girl opened the note and said: "I am so sorry. He has been called urgently to Mexico City. He went on the midnight train. He doesn't say when he will be back."

Borlaug knew he would never come back; his slick operation had been cut short. There was still time to offset its effect before his work was damaged beyond repair. He formed his team into a group to spread the warning. They ranged out along the valley, asking people if they had ordered the fertilizers. How had they paid? Could they order the banks to stop payment on the checks?

Don Rodolfo had already received a hundred tons and was about to spread it on his land. It was hard to convince him that it was wasted, useless. Some were luckier and could halt their bank drafts; others had been bitten.

Borlaug reviewed the situation with his colleagues at the Yaqui station. When each man had reported on his findings, he said: "It seems we were in time to avoid total setback with the major farmers; but with the smaller men it looks like a disaster that will take us at least three years to overcome."

They did not see it then, but the setback held a silver lining. It pointed up the need for the farmers' own research and scientific center, staffed by reputable scientists whose word could be trusted.

About a month later Roberto Maurer and Don Rodolfo drove out to see Borlaug at the Yaqui station. All the big farmers had met, they told him, to consider Don Rodolfo's proposal that they establish a new research institute in the valley.

"It is clear we need a center close to us," said Don Rodolfo, "which will deal with our problems, which will work for all our futures, and which will give us good men all the year round."

Not only would it be good for the valley; it would be good for the entire state of Sonora. He had enlisted support on that basis.

"We know a piece of land—250 acres—to which we can add later. It is closer to Ciudad Obregón, and that is important for the young men who will come and work with you. They will want to send their children to school and have their homes near the shops."

Don Rodolfo said he had heard about the land from a "friend in federal cabinet." It was ideal for a research station, with plenty of irrigation water, and the price was reasonable. With buildings, laboratories, and stores it could be put into operation for a hundred thousand dollars.

The two men told him the center would be for him to use and operate as he wished. It could be incorporated into the Office project, and the farmers would pay a levy on their wheat crops toward the maintenance. It was done quietly and quickly. The land was acquired, the architects were brought down from California, and the project was under way in two months. The new center eventually became known as CIANO—Centro de Investigaciones Agrícolas del Noroeste. It was the northern outpost of the Mexican project and eventually became the regional station of the Mexican organization that took over research and training from the Rockefeller Foundation. Scientists from many nations would come there to study, and local wheat growers would pay $400,000 each year to keep it operating—five times the sum given by the federal and state governments combined.

Though Borlaug was spending time in the Yaqui, he had not been neglecting his research. There was, in fact, one particular development at the Chapingo station that he was watching with increasing interest.

In the summer of 1953 he had received a small packet of seeds from Dr. Orville A. Vogel in the United States with a note saying that Dr. B. B. Bayles had suggested he might be interested in them. They were second-generation seeds of a cross Vogel had made between an original dwarf-type wheat from Japan and a tall wheat from North America called Brevor. These were the wheats Borlaug had discussed with Bayles on their South American trip, and he decided to plant them immediately. They were a winter variety, late headers, and highly susceptible to disease, but he was interested in them for their short, strong straw.

For uncounted centuries these so-called dwarf wheats had clung to the soils of Asia. They were descended from wild *triticum* plants, precursors of the domesticated wheat that had spread across the globe with man, adapting to various environmental pressures. In the Western hemisphere, where conditions were favorable, centuries of cultivation had produced wheats with straight, tall straws. But in Asia—in the plateau valleys of Tibet, in the reaches of northern China, in Mongolia, along the vast sweep of southeast Russia—the genes and chromosomes had struggled to adapt to short summers, severe winters, and savage, battering winds. The result was a hardy wheat of strong, short straws, with more stalks per plant and consequently more grain.

Western science had long known about the Asian dwarf wheats, but it was not until the end of World War II that they made any real impact. When Japan surrendered in 1945 and General Douglas MacArthur went into Tokyo with an occupation force, he took with him a team of specialists to help with Japanese reconstruction. Among them was Dr. S. C. Salmon, an agronomist with the U.S. Department of Agriculture. During his travels through Japan Dr. Salmon visited the Morioka experimental station in northern Honshu and was surprised to see wheat growing like little bushes about two feet tall. It seemed to be compensating for its lack of height by tillering, thrusting up more stems from the base of the plant. Upon inquiry, Salmon found that the little wheat was called Norin 10, a descendant of some strain that had long ago crossed the China Sea, and that it responded very quickly to water and nutrient.

Intrigued by the possibilities of Norin, he gathered seeds from a dozen or so plants and took them back to the United States, where he gave them to Dr. Orville Vogel for closer study. Vogel put the seeds into quarantine and grew them in detention to check whether they carried any diseases new to North America. He established that they were a winter variety, extremely susceptible to leaf stripe and powdery mildew.

Having cleared the decks of any danger, he began crossing the Norin with two American wheats, Brevor and Baart. It was seed from the second generation of the Norin-Brevor crosses that he sent to Borlaug in Mexico.

The wheat grew well in Borlaug's experimental plot at Chapingo, and late in April 1954 the sixty little bush plants were close to pollination. It was late for most of his other varieties, and the only line of plants that offered him pollen for crossing was a new wheat called Yaktana, a cross between Yaqui and Mentana produced by Borlaug's former assistant, Joe Rupert, in Colombia.

He set to work with Alfredo Campos emasculating the little anther stamens from the Norin dwarfs. The task was more painstaking and delicate than usual, because the plants were so small. When they had all been rendered female, Borlaug and Campos selected heads from the Yaktana and dispersed the pollen inside the small envelopes.

It was disastrous. The male heads of the Yaktana were impervious to rust themselves, but rust spores had settled on them and were dispersed with the pollen into the Norin heads. The fungus devastated the little nursery of dwarfs. Every one of them died.

When they saw the results, they were aghast. "We should have made the dwarfs male and this wouldn't have happened," said Campos.

Borlaug knew he was right, but it was wisdom in hindsight. On a hunch he went to the shed where he had left the envelope containing the packet of seeds and shook it into the palm of his hand. Eight seeds fell out. Borlaug looked at Campos and smiled. "It's enough, Alfredo. In the final analysis we need only one seed—if it's the right seed. We'll take good care of these."

That summer Borlaug spent as much time as possible with Margaret and the children. Billy's health continued to be a problem, and Margaret thought he should be encouraged in physical exercise of some sort to build his strength. Borlaug discussed the matter with his colleague John Niederhauser, who lived nearby and had two sons close to Billy's age. Niederhauser had heard about the growing popularity of Little League baseball back in the States, and Borlaug embraced the idea at once. He regarded the creation of a competitive spirit and a team outlook vital to the development of character in a young boy. They had no trouble finding ballplayers, and together they lined up parents to alternate responsibility for coaching, training, and organizing a league.

The interest created a closer bond between father and son. Borlaug became an enthusiastic baseball coach. His participation continued into the late 1950s and early 1960s, as he took new teams under his wing, some composed of young Mexican boys who showed spirit and promise.

He helped to organize the first Mexican Little League and the first all-Mexican tournaments. It gave him many happy weekend hours with young people and helped to relieve the tensions of his ever-increasing workload.

Ten

In the fall of 1954 Norman Borlaug allowed himself the luxury of good spirits. His work, he had reason to feel, had achieved a new status. From now on his experiments would be conducted as they were meant to be conducted—in proper facilities under properly scientific conditions. And he had more to work with now, a greater variety of wheats, plus closer connections with the international scientific community and a growing corps of well-trained, dedicated young Mexican scientists.

Some evidence of his new status was contained in the score of seed packets that he took from Chapingo to the Yaqui. Never had he had so great a variety of seeds, strains sent to him from agronomists around the world. In scholarly journals during 1953 and 1954, Borlaug and his associates had published a number of articles that had attracted international interest. He had also shared authorship of several papers with Stakman, Bayles, and Joe Rupert and with his Mexican colleagues Ignacio Narváez and Alfredo Campos. In his name alone Borlaug had published several reports in agricultural journals in the United States and Canada, and he had described the new era in Mexican wheat production. Not surprisingly, as all of this written material made Borlaug and the others familiar to the agricultural-scientific community, packets of seeds had begun to come into the Office. Some of the senders wanted the seeds tested; others wanted them to be crossed with the best of the antirust wheats among the Mexican giants. Thus the Yaqui and Chapingo stations had begun to serve as worldwide growth-testing centers for wheat, and as clearinghouses for the free exchange of data.

More evidence of the emergence of better times was the new CIANO complex at the Yaqui Valley station. Don Rodolfo's extraordinary powers of persuasion had worked wonders. Borlaug was astonished by the enthusiasm of the Mexican workers toiling to complete the quarters and further astonished by the state of near-readiness of the facilities. The fields still had to be leveled and sectioned into experimental plots, and the water supply system was not yet completed; but otherwise the premises were ready to function. Rooms and offices were waiting to be put to use, and laboratories were sufficiently equipped to operate. No wonder Borlaug felt renewed: the days of hardship and make-do were behind him.

In one of the new CIANO buildings, a plant protection laboratory, Borlaug set aside a corner where tender, susceptible seedlings could be coddled indoors and protected from wind and rust attacks. It was here that the eight tiny, wrinkled seeds of the Norin-Brevor dwarfs would be grown. Because of their origin and their undefined but anticipated promise, they needed special treatment. They were winter wheats, accustomed to the cold seasons of Japan, and Borlaug wanted to simulate their familiar environment. He placed them in a dish on a bed of cotton wool that had been soaked in nutrient and put the dish in the refrigerator for three weeks. There was nothing new about this procedure, which plant breeders know as vernalization. Seeds with ancestry in cold climates grow better when, prior to planting, they are subjected to a cold shock. (The Eskimos practice a sort of vernalization on newborn babies, who are bathed in snow immediately upon delivery.) Release from the cold triggers the genetic responses in the wheat germ plasm in the kernels.

After the seeds were taken from the refrigerator, Borlaug planted them in pots and set up muslin screens behind them to protect them from the rust spores that might float into the laboratory on air currents. Not until the genes of rust resistance in his Mexican giants were bred into these foreign strains would he risk an outdoor planting.

Furthermore, the mating role of the dwarfs was to be changed from female to male. Borlaug planned to introduce their pollen to selected rust-resistant lines in his outdoor plots. With their male stamens removed, the Mexican giants would become female: if any rust spores were introduced, the plants would be able to defy the disease. For the crossing Borlaug had in mind the best new antirust wheats in the nurseries, including fresh lines bred with combinations of the Hope and Kenya defenses, a line known as Lerma Rojo, and the new Yaqui and Kentana wheats. Of all the lines developed in that decade Lerma Rojo would prove to be the wheat whose rust-resistant property lasted longest, and its offspring would provide a necessary ingredient for the coming of the Green Revolution.

The eight dwarfs remained in isolation in a corner of the laboratory throughout the Yaqui winter. The little seeds sprouted and grew steadily.

Borlaug had more to do than watch them grow. Among his activities was a meeting in Ciudad Obregón with the two farmer groups—the association formed by the ejido farmers and the cooperative. To both he continued to hammer home his lessons about fertilizing for increased yield. Production for the nation for the coming season had been officially estimated at 850,000 tons of wheat grain—double the figure of ten years before. Dissatisfied with this projection, Borlaug told the farmers: "You have got to come to your senses! If you irrigate your wheat fields and you do not give that land a feed of fertilizer, then you simply waste water—and that is unpatriotic in Mexico." The farmers had heard this from Borlaug before. Ever since he had taken over the project he had exhorted them to apply his ideas: wider use of fertilizer; controlled use of irrigation water; increased yield potential in varieties of wheat; better defenses against rust. And they had listened. Under Borlaug's directorship of the project Mexico had doubled its wheat grain harvest, and in 1955 it was only a year away from self-sufficiency in grain production. Nevertheless, certain that an even greater yield was possible, Borlaug continued to press for more vigorous application of his program.

He also had some good news for the farmers. An interesting development had occurred and Borlaug was jubilant about its implications.

For a decade he had moved wheat strains back and forth twice a year between the sea and mountain plots. The annual environment reversal had put pressure on the genetic material in the wheat germ plasm to adapt. Adaptation, of course, is fundamental to all forms of life, and most plants are photosensitive. Inasmuch as plants absorb much of their nutrition and generate the energy necessary for growth from sunlight, they are equipped with a biochemical "camera" that measures and records the amount of sunlight available each day. The mechanism is triggered by an enzyme produced by the genes. When the camera discovers that the sunlight supply is consistently insufficient to nourish the plant *and* to provide for continued growth, it informs the genes, which curtail the growing process. Thus during the winter months, when sunlight is in shortest supply, most plants, merely to stay alive and healthy, stop growing in order to conserve the sun's nutrients.

Because they were moved back and forth twice a year to locations where daylight was in ample supply, Borlaug's selections were always exposed to light in excess of the minimum needed to generate growth. Thus their growth mechanism was never turned off. And after several generations of selecting those individual plants which were not using their photosensitive mechanism, he noticed some selections grew equally well in Chapingo or Yaqui. When an entity is no longer needed, it ceases to be operative (evolution is a process of elimination as well as of adaptive development). In Borlaug's wheats the genes that produced the enzyme that triggered the camera stopped operating. The Mexican wheats had become photosensitive.

It was a fundamental advance, and Borlaug was jubilant. His wheats were now oblivious to the length of the day. The world would receive a migration of new wheat strains.

At Ciudad Obregón he discussed the development with Narváez, Campos, and other Mexican scientists: "There are going to be plenty of people who won't want to believe what we have done. I have wondered for years why wheats from Canada, Montana, Minnesota, Dakota, and other places could not do well down here, outside their homelands—why they grew so well in their home soils but were miserable and unhappy when they were moved over any distances and became poorly adapted, low yielders. What happened with them was that the damn biochemical camera was alerting the genes to switch off the growth enzymes. It won't happen any more. When the plants can't read the number of daylight hours, we can move these varieties far and wide across the world."

In the early weeks of 1955 the Norin-Brevor dwarfs in the CIANO laboratory caught and held Borlaug's attention.

Properly fed, the little bush plants proved that they could tiller strongly—that is, produce an ever-increasing number of stems on which seed heads would grow. Moreover, each head had more grains than ordinary wheats. These were immeasurably valuable attributes; combined with the most desirable characteristics of the Mexican giants, it could create—what? A dwarf plant with good yields that was rust-resistant and insensitive to the length of day. But could it be done? Could he merge the genes that would unite all the vital traits in a single plant line? If he could, Borlaug knew he would have that golden nugget from which would grow an ideal wheat.

When the heads of the dwarfs were close to maturity and the pollen anthers near to bursting, Borlaug and two colleagues, Dr. Alfredo Campos and Dr. John Gibler, prepared the female plants. They chose eight lines of the best anti-rust giants, carefully plucked out all the male stamens, and covered the emasculated plant heads with paper envelopes. By then the flowers on the dwarf spikes were open; the pollen was ready and waiting.

Borlaug cut the heads off his eight little bush plants and carried them to the wheats waiting in the outdoor nurseries. The spikes were inserted into the white envelopes and twirled around to disperse the pollen. Then came the waiting. Weeks would pass before Borlaug would know if the crosses had been made, if he would have a new seed to plant in the summer at Chapingo. Even then he could not be sure that a seed produced by the mating of an Asian dwarf and a Mexican giant would emerge fertile. There was always the chance that it would produce a mule—a variety from which he could not breed properly. It was a case of wait and see, then replant and wait some more.

In 1956, for the first time in the history of the republic, Mexico fed itself. Mexican farmers produced 1.25 million tons of grain—more than enough for self-sufficiency. Indeed, the supply of wheat was large enough to offset a shortage of corn. The Rockefeller project—which had been expanded considerably with the addition of more soil experts and plant breeders in corn, beans, and potatoes—had led Mexico to its agricultural independence in a period of just thirteen years. And in that time—short in agronomic terms—Borlaug's originality, perseverance, and drive had resulted in the development of dozens of new varieties of wheat.

Borlaug's Mexican nurseries abounded with some 50,000 varieties and hybrid lines of wheat in 1956. It certainly looked as though Mexico had won its wheat revolution. Money that formerly would have been spent abroad for grain now went into the hands of farmers, campesinos, and the little ejido families. True, costs had doubled in six years: the price of farm machinery and tools, fuel, fertilizer, and labor had increased by more than 60 percent. But the price of wheat had risen by only 10 percent, and yields had been tripled—at the very least. Farmers in the Bajío were spending 250 pesos per acre for fertilizer alone—a figure that would have been unthinkable ten years earlier. But the same acre that had yielded six to ten bushels of grain in 1946 was producing between forty-six and sixty bushels in 1956.

The event drew excited comment from the Mexican press. The leading Mexico City newspaper headlines proclaimed "More Wheat This Year" and told of record harvests. The magazine *Tierra* said that "agricultural revolution is no more a figure of speech. . . . Mexico is on the road of tremendous progress." The magazine further pointed out that the achievements "serve to refute the unfounded charges by the detractors of agricultural scientists . . . that agrónomos engage in everything but the practice of their profession."

The new activity in food production and the new presence of money in agrarian hands stimulated the growth of supporting industries. Demand for new farm equipment and supplies increased; so did the market for small luxuries, from radios to bicycles to better clothing. Not only did the achievement of self-sufficiency provide jobs and lift the economy—more than 60 percent of the Mexican population lived from the new, productive soils—but it created a growing corps of young Mexican scientists. Naturally within that corps there arose the requisite Young Turks, determined to establish national independence in the scientific as well as the practical branches of agriculture. But that was all right with the original sponsors of the project. The growth of the nationalistic scientific movement only brought the Rockefeller gringos closer to the objective set for them by Harrar in 1943: "to work themselves out of a job, as quickly as possible."

Borlaug could not possibly supply the demand for one product: himself. The government of Chile wanted him to head its projected rural revolution. Resisting strong pressure and attractive offers, Borlaug recommended his Rockefeller colleague Dr. Joseph Rupert to the Chileans, and his suggestion was accepted. Several American universities made it known that a comfortable, lucrative, influential position could be Borlaug's if he was interested. He was not interested. But he did seriously consider one offer—from the banana industry. He was mainly attracted by the challenge of its problem. A fungus known as Panama disease was spreading, and the banana producers thought that Borlaug was the man to defeat it. Moreover, the offered stipend of $18,000 was double his Rockefeller salary. There was no doubt that he was tempted, and negotiations went on for many months.

But the length of negotiations worked against the banana industry. As the initial excitement generated by reaching the goal of self-sufficiency in Mexico quieted, Borlaug's thoughts returned, more and more often, to the experiment he had begun in the Yaqui nursery—his attempt to cross Mexican wheats with dwarf plants from Asia. Against it the bananas did not stand a chance.

Borlaug visited the strange little plants every day. As the weeks went by, new shoots and stems kept popping out on the plants. The strong tillering characteristic of the dwarf seemed to have been successfully inherited by the seed. As it thrust up from the soil, each new stem carried more leaves, closer together. The new plants created problems and revealed deficiencies. But they also filled Borlaug with hope.

Ignacio Narváez received his doctorate from Purdue in 1957 and returned to Mexico to work with Borlaug on the new dwarf wheat project. In all his years with Borlaug Narváez had never seen his teacher concentrate with such intensity on a single variety. He watched Borlaug's mood change as he studied the dwarfs—now angry, now delighted, now frustrated, but always, always patient. Month after month Narváez saw Borlaug shaking a problem to pieces like a terrier with a knotted rag, worrying and tossing it back and forth, never letting it go. Slowly, as the two men worked through the second and third generations of dwarfs, they began to put pieces together. It was like assembling an enormously complicated jigsaw puzzle with no model on the boxtop to work from.

The model was a dream—a wheat different from anything the world had ever known.

In the 1957–1958 winter season Borlaug took Billy with him to the Yaqui for the boy's first visit to his father's new research center. Billy was ten and thought the holiday in the northwest a high adventure. Borlaug enjoyed his son's company, but once work got under way he saw little of him.

They spent a few days together in the fields, though for the most part Billy passed the time playing with the children of Roberto and Teresa Maurer. The experience left Borlaug with a feeling of discomfort. For a long time he had been hoping that his son would share his enthusiasm for his work and eventually make a career for himself in plant breeding. Billy never did take any special interest in biology and later decided to study economics.

While Billy was there, Dr. Stakman arrived for a visit. Now retired, Stakman had retained his deep, almost paternal affection for Borlaug and his family, and he had come to make a plea: he wanted Borlaug to leave the project, or at the very least to take a long break from it. The aging scientist was concerned that the long separations between Borlaug and Margaret, alone in her suburban Mexico City apartment, were harmful to them both.

But the idea of abandoning his work on the new semi-dwarf wheats was intolerable to Borlaug. Moreover, he pointed out, Mexico would soon need the new strains. Wheat production had reached a ceiling: no more wheat could be produced than was being produced. And, Borlaug reminded Stakman, in the thirteen years it had taken them to develop the forty new varieties of wheat that made Mexico self-sufficient, Mexican families had produced 10 million babies. Should births continue at that rate, it would not be long before Mexico returned to the underproductive poverty level of 1943. To make matters worse, ten new virulent rust races had appeared, and it was not at all certain that the existing wheat strains would prove to be impervious to infection. If he could establish the correct genetic breeding patterns, Borlaug was sure that the new wheat would provide protection against the gathering problems. "I know there are drawbacks and defects," Borlaug told Stakman, "but I feel suddenly we are in business again! These are really interesting. We get some plants with soft seeds—no good for making bread; we get defective heads shooting too early from the soil; we get partially sterile heads. But we are also getting heads with little spikelets that carry six grains in place of the normal three. We are getting tillering, more and more spikes, more heads."

Clearly Borlaug was not going to consider leaving while his experiment was progressing so steadily.

He showed Stakman the new types. The bushes had a golden, fluffy look with heavily grained heads. They grew only knee-high, and there already were several different variations, according to the parent strains of the Mexican wheats. They would come out all right, he declared; he knew they would. He was expressing this confidence when Stakman told him that the way things were developing Borlaug might not have any choice in the matter: there was a crisis with the Mexican government, and if the political ax fell it might very well be directed to the Rockefeller neck.

The gringos who had come to Mexico to work on the Rockefeller project had met a certain suspicion and resentment from the start. The more dedicated among them were able to overcome the prejudice of the Mexicans merely by doing their jobs well and by cooperating with Mexican scientists and farmers, but not all were that dedicated or diplomatic. Especially after the achievement of self-sufficiency, some of the Rockefeller people tended to speak as though the achievement were theirs, ignoring the contributions of the Mexican agrónomos, farmers, and government. The Young Turks in the Mexican scientific community were complaining. How long, they wanted to know, were the Rockefeller people going to remain in Mexico and run the show? Now, Stakman said, the young administrative scientists had taken their case to the minister of agriculture, General Gilberto Flores Menôz. A fiercely patriotic soldier politician, General Menôz had quite a reputation. He had ended many careers among civil servants and government contract workers over the years with his famous, to-the-point telegrams: "Cessado [dismissed], Menôz." He gave no explanations; no one asked for any, and no one argued. Apparently, Stakman concluded, Menôz had taken up the cause of the Young Turks, and there was a very good chance that the Rockefeller Foundation team would be dismissed from Mexico.

Leaving Borlaug with enough to think about, Stakman said good-by and returned to Mexico City. After his departure Borlaug drove out to see Roberto Maurer. Together they called on Don Rodolfo Calles. Borlaug was emphatic: "If we don't solve this problem our work will be stopped. We'll get a one-word telegram from Menôz and it will mean we leave Mexico—for good." Borlaug's Mexican colleagues agreed to try to find out what they could, and the next day both confirmed the worst suspicions. The Rockefeller personnel were about to be deported. Don Rodolfo, however, had arranged for a special delegation to meet with General Menôz. He suggested that Roberto appear at the interview along with another officer of the Yaqui farmers' cooperative and a representative of the ejidos' association. The three-man team was assembled and flew to Mexico City that day.

The very fact that the trip was necessary angered Roberto Maurer. Getting rid of the Rockefeller people would be folly, he thought, and the idea had sprung from the ambition of self-seeking men. He arrived at General Menôz's office, and the general's demeanor—cool, smooth, soft-spoken, and even friendly—angered him more. Of course Menôz would be friendly: the farmers' vote was important. Roberto spoke boldly and scathingly. If the minister of agriculture carried out his plan it would be a gross injustice, based on lies and deception. The ejidos' representative pulled Roberto's sleeve and interrupted his harangue. "Be careful my friend. They are powerful men. They can do you much harm in the future."

Menôz admitted that he was about to banish the Rockefeller Foundation from Mexico. His decision was based on certain evidence that had been submitted to him. If, however, the farmers' representatives had contrary evidence, he would like to see it; he might be persuaded to change his mind. Roberto said he certainly had evidence. (Don Rodolfo had anticipated the request and secured the necessary documents before the delegation had left the Yaqui.) The general sent for three agriculture ministry officials. Menôz did not introduce them, and Roberto recognized none. The general asked them to state the false claims made by the Rockefeller people. The officials recited a list of recently developed varieties of corn, wheat, and beans that had been claimed by the Rockefeller group but that, they said, actually had been grown and bred by Mexicans on Mexican stations. On the list were several wheats that had been bred and named at Chapingo and in the Yaqui.

Roberto rebutted with a sworn statement, signed by himself and Don Rodolfo Calles, detailing the pedigree wheats, with all their antecedents, that Borlaug and his Mexican trainees had produced. General Menôz read the statement and handed it to the officials without a word. He asked his question silently—with raised eyebrows. Silently he was answered—with confounded expressions. With a martial, imperious wave of his hand, the general dismissed the officials.

As he shook hands with Roberto and his companions, General Menôz said: "They will be dealt with—those people—in a most proper fashion, señors, you may be certain of that. But you were lucky. You were just in time." He tapped a document sitting on his desk. "Once I had signed this paper, I would have fought you to the end before I would have changed my mind."

In Mexico City neither Wellhausen nor Stakman knew of the meeting between General Menôz and the delegation from the Sonora farmers' groups. They went ahead with plans for their own confrontation with the minister of agriculture. So important did they regard this meeting that Stakman contacted New York and suggested that the Rockefeller Foundation president, Dean Rusk, attend. Rusk could not get away but sent one of his vice presidents, Lindsley F. Kimball, and another senior officer, Kenneth Wernimont, to join Wellhausen and Stakman.

Although the meeting had a happy ending, it did not go as smoothly as the one between Menôz and the farmers' representatives (which, for some reason, Menôz never mentioned). Perhaps the general had been more impressed with the humbler delegation and with Roberto's conspicuous anger than with the diplomatic bearing of the high-echelon Rockefeller spokesmen. Whatever the reason, Menôz launched a tirade of abuse against the Rockefeller activities in Mexico, his chief allegations being that the foundation was binding Mexican farming too tightly to American interests and making Mexican agriculture entirely too dependent on American good will. In general terms there was some validity to Menôz's accusations, but apart from his complaint that American scientists had claimed too much of the credit for the success of Mexico's agrarian revolution, the specific examples the general offered were not true. Stakman said flatly that Menôz had not been getting accurate information but then admitted: "No one is without fault. The Bible tells you that. We are not perfect. Who is? We have given many years of work and effort and money for Mexico. We make no profit. We seek to help your people to get the very best life possible. We admit to being gringos, but we try, most of us, to behave ourselves in your country. We do no harm—only good. There are many countries—some of them in South America—that are waiting for Rockefeller assistance impatiently, that want our science, our skills, and our help. If you don't want us to expand and develop what we have done so far, then say it in a friendly way. We will shake hands and go."

Stakman's speech impressed General Menôz, who stopped his attacks and admitted that Mexico needed more help, not less, from Rockefeller—if the help came without strings. Specifically, the general asked that a sort of liaison service be established. One arm of the service would keep the government informed of the work of the foundation; the other arm would see to it that the ordinary farmer was kept abreast of the latest scientific developments and helped to put the developments into practice. The Rockefeller delegates said they thought such a service could be established —and a grant of $60,000 per year provided for its operation—if the general could assure them that the foundation's projects would continue without swords over their heads. The general gave his assurance, admitted that he may have been misled, and took them all to dinner.

Yet the whole episode caused the Rockefeller Foundation to reappraise its objectives. The problems with the Mexican scientists and General Menôz pointed up the fact that the foundation's projects were most effective at the scientific level; trouble came after the scientific objectives had been achieved, at the administrative level. From now on the foundation would reduce its administrative participation in projects that already had succeeded. Now there was a bigger task to tackle. Hunger would be fought on a global scale.

In 1959 Borlaug was informed of the coming changes and was asked to undertake a survey mission in South America to suggest areas to which the Rockefeller Foundation could extend its work. It was a long, tiring journey that took him to Brazil, Argentina, Bolivia, Peru, Ecuador, and Chile. He was distressed but not surprised to find the same conditions almost everywhere he went: plenty of natural resources waiting to be tapped but wasted in the meantime; lack of activity, initiative, and elementary knowledge at the farm level; native scientists with degrees from overseas universities—mostly the United States and Europe—falling into the familiar pattern of academic isolation: men in white smocks wrapped in their own security, aloof from poverty and hunger and the realities of the land. Particularly astonishing to Borlaug was the fact that so many of the scientists either held or were pursuing degrees in fields completely unrelated to the great problems of their homelands.

In the following year the news came that the Office was to be closed at the end of 1960 and its responsibilities taken over by Mexico's new National Institute for Agricultural Research. A new Rockefeller organization, based in Mexico, would be established: the Inter-American Food Crop Improvement Program. Within its structure would be three international subprograms. Wellhausen would direct the corn improvement project, Borlaug would head the wheat program, and Niederhauser would take charge of the potato project.

In January 1960 Borlaug was asked to name his successor at CIANO; the designee would take over in February. It was hard to turn over the reins. How could he detach himself from all that he had put in motion? How could he leave his work in the Yaqui? He was so distressed that Wellhausen spoke to the Mexican authorities and returned with good news: Borlaug would be regarded as counselor and collaborator and would have unlimited use of CIANO. Inasmuch as he would be based in Mexico, he could reach the familiar nursery plots at Chapingo and in the Yaqui Valley fairly easily.

With that settled, Borlaug turned to the designation of a successor. Without hesitation he recommended that wheat research in Mexico be placed in the capable hands of a man he had first known as a strong-spirited agrónomo caught in a flooded mountain gully beyond Saltillo: Dr. Ignacio Narváez.

Borlaug was saddened that year to learn that Mrs. Jones had died. The coffee and cake sessions would be no more; he would miss her friendship and good counsel. She had been the last Yankee of the pioneer settlement days in the Yaqui. Her two daughters did not return to take over the land, which was sold to another Yaqui farmer. The old Jones home was pulled down. Eventually dwarf wheats were to cover the Jones acreage. Borlaug would always remember his friend with gratitude and affection.

He and Narváez paid frequent visits to the semidwarf wheats. It was thrilling to see how the yield and quality were improving. The strain was not yet ready for multiplication: it still had soft centers, which gave it a poor breadmaking capability. No, the dwarfs were not the best that could be produced. But they were improving. It was only a matter of time.

At Christmas Borlaug went home to Mexico City. During his holiday Harrar called from New York. The Food and Agricultural Organization of the United Nations had asked the Rockefeller Foundation to make Borlaug available for a fact-finding survey mission to the Near and Middle East. Borlaug said that if he could be of help to the world body he was ready to go.

In the course of his work he had traveled north and south. Now, for the first time, he was packing to go east.

Eleven

The fact finders met in Rome early in February 1960. Among them were several Food and Agricultural Organization officials, notably Dr. Abdul Hafiz of Pakistan. Dr. James Harrington of Canada, an adviser to the FAO wheat and barley program, was there, and so was Dr. José Vallega of Argentina, a biologist who was about to join the FAO as a full-time officer. Borlaug was the "man from Rockefeller"; his presence was important because the foundation and the FAO were about to merge their efforts in a program to deal with food shortages worldwide.

The specific mission of the party that gathered in Rome was to inspect the progress of the FAO wheat and barley program, which was already under way from the Near and Middle East to the Indian subcontinent. India was not a participant in the project; but the Rockefeller Foundation had established an office in New Delhi to study India's staggering food problems, so the country was included in the itinerary of the UN survey mission.

Under the UN flag the scientists and administrators traveled by air and over land across North Africa—Algeria, Libya, Egypt. Then they made stops in Jordan, Lebanon, Cyprus, and—after scratching Iraq from their schedule because of internal unrest—Iran. Afghanistan, Pakistan, and India completed their tour.

At experimental stations in many lands Borlaug saw plots of his Mexican-bred wheats. "These are brothers and sisters of my wheats at home," he told his colleagues. "I know them by their first names."

But there was little to feel familiar with in India, where the food problems were the most severe. India was where people were hungriest, where the soil and the elements seemed most inhospitable to the needs of agriculture, where the very social system seemed to work against scientific progress. Borlaug knew how bad things were; he had seen the data, read the reports prepared by the foundation's New Delhi office. But seeing was something else.

In 1952 India had asked the Rockefeller Foundation for help in overcoming its grain food problems. At the time rice was the main food crop. The annual rice yield was about 35 million tons—more than twice that of corn, millet, and sorghum and five times the annual harvest of wheat. A foundation team that included Harrar, Warren Weaver, Richard Bradfield, and Paul Mangelsdorf responded promptly and conducted a thorough investigation of the country's agricultural output and problems. Its report formed the basis for a set of proposals, prepared by Wellhausen and a Rockefeller colleague, Dr. U. J. Grant, for increasing the output and correcting the problems. They recommended many of the same steps that had been taken in the Mexican project—the injection of new plant materials in corn, the launching of an objective research project coordinated on a national level, and the reorganization and redirection of educational facilities. The foundation appointed Dr. Ralph W. Cummings to effect the essential changes in education. Under Dr. Cummings, the Pusa Institute in New Delhi became the Indian Agricultural Research Institute—IARI—which trained some of the finest scientists in the land. Later Dr. Cummings would be appointed chairman of the Indian Agricultural Universities Committee, which meant that he was virtually the dean of all domestically trained agricultural scientists.

Yet the food situation in India improved only grudgingly. India was the home of one-sixth of the world's population; its numbers were increasing very rapidly, but its food supplies were increasing very slowly. The scientists who graduated each year from India's fifty agriculture colleges did not seem to help much.

It was Borlaug's opinion that the scientists, still trained along British lines, lacked direction and motivation. As far as he was concerned their training was a waste. The pattern he had seen in Mexico and South America was duplicated in India. Graduates had no sense of realism. Degrees were awarded on the basis of theoretical rather than practical achievement. The dislike of scientists for working in the fields, which had been so bad in Mexico, was endemic in India. The standing of the agricultural scientist was very low with common people and politicians alike. Industrialization was a bright bauble; nuclear power a golden dream. Food production was way down on the list.

The canker went beyond this. Cummings told Borlaug that most professors held identical views. There was still a strange mystique in the land, far removed from reality; and many were oblivious to the fact that only gift food from abroad saved India's millions from starvation. Impoverishment was more degrading and disheartening in India than in any country on earth, and that, in Borlaug's mind, made the apathy and acceptance of the situation unbearably stupid. His exasperation and anger increased when he realized that India was not so poor as it seemed. It had the natural resources with which to climb out of its lethargy. But the resources were untapped, ignored.

Throughout recorded history the people of the Indian

subcontinent have been the hungriest on earth. Famine has been commonplace. Droughts could be counted on, one or two a decade, and when they came people fell dead in uncounted numbers. No wonder that life became cheap. The soils of India were poor, baked by the savage sun and overworked, but the problem was mainly lack of water. Yet for all of human time the mighty mountain range of the Himalayas, which stood to the north of this aching, dry country, had trapped colossal amounts of moisture from the atmosphere. Enough water flowed from those elevated snows to cover the subcontinent to a depth of twenty inches each year! But beyond the rivers and the few irrigation areas the government had done little.

Worse, the flow of new babies ran faster than the great rivers, while the land area to feed the population had steadily decreased from two acres to two-thirds of an acre per capita, and it was still dwindling. The pressure on the poorest of the poor was crushing. Men, women and children died in the country roads and in the city streets. The appalling fact to Borlaug's mind was the fatalism, the acceptance, and the disregard for the misery of fellow human beings. Years later the glaring inhumanity was still blazoned in his mind, the details still very clear, when he said: "They all stuck stubbornly to their ancient methods of agriculture. They seemed not to think about it. The way they took this enormous poverty clobbered my mind. It weighed down most of all on that poor, skinny farmer and his large family. But it did not seem to affect the many nonfunctioning scientists. This shocked me. In some of these countries they had the excuse of no corps of science for agriculture. But in India, where they had many hundreds of scientists, new ideas—initiation of development—were suspected and rejected. When you tried to find out why nothing much was happening, you came up against a smothering wall of bureaucracy.

"I looked around there and talked with these men, and what I learned was later confirmed with emphasis. Every one of them was digging his little gopher hole of security in his own discipline, becoming entrenched against removal from the job, obsessed with the status and security it brought him and his family, and never lifting his head to look over the city wall to see what was going on outside.

"Here was the pattern on the Mexican blanket all over again, thread for thread: impoverished soils, lack of science, a backward rural economy, a people wallowing in poverty of capital and ideas and totally lacking in leadership—but it was on a much vaster scale."

Norman Borlaug was always happiest when he was close to the land. In Mexico he had been able to unroll his sleeping bag on the ground or on a rough cot, climb in, and get a good night's sleep. He had never managed, however, to sleep in an airplane, and he had to accustom himself to arriving in each new country tired and bleary-eyed.

Similarly, when he returned to Rome after his eastward trip, he found it difficult to begin writing his report. The environment wasn't right. He was supposed to sit down and propose methods to relieve the food shortages over the vast region he had traveled, to write out a plan to feed a billion of the hungriest, saddest, poorest, most desperate yet lethargic people on earth. And in a stylish Roman hotel, with bath and shower and room service, it was difficult. He could more easily write the report in his first cottage in the Yaqui.

He left the hotel and walked through the cool, sunny streets; animated, purposeful faces contrasted sharply with the anxious, hopeless faces of the Indian people and their timid children with big, staring black eyes. And whenever he heard Roman laughter, he remembered how rarely he had seen a child laugh in India or Pakistan.

He walked across the square to St. Peter's, went in, and sat against a far wall, his chin on his chest, thinking. What could he suggest? Should he argue that the world should marshal forces against this great sweep of human hunger and misery? Lord Boyd Orr, the craggy Scot who led the FAO, had warned again and again that the food bins of the planet were fast emptying, that the United Nations should rally for a massive global plan to attack famine. He had run aground on international politics. Half of humanity went to bed hungry, and the other half did not care. That was the situation. What was the cure? What relevance did Mexico have to this pattern? What relationship could be established? It had taken all of sixteen years in Mexico, but India, Pakistan, the Middle East, and North Africa could not wait that long without risking enormous suffering. Many of those lands had few trained people. Afghanistan had none. India had many—but few with motivation. Libya had half a dozen men with training; Algeria only two or three. Cyprus, Iran, Pakistan—all needed young blood.

It was perfectly natural sitting there in St. Peter's that he should think of the word he was looking for. There would have to be a new breed of apostles—agricultural apostles. Back in the hotel room he began writing:

"Lack of money means lack of food, and the reverse is also the case. One hangs on the other in lands where the rural people form three-quarters or more of the populace. Yet to inject aggressive technology, modern machinery, and methods into this primitive industry would require massive capital and burden future generations with loan repayments. On those hundreds of millions of rundown acres, scratched over by animal and human muscle power, in an industry dominated by the ox and the wooden plow, only new confidence and hope among the people can bring revival. It is all beyond their governments and their governments' resources. I have heard the fatal excuse—as I have in the other struggling countries of Central and South America—that the peasants want no more than their fathers and grandfathers had—that they do not want change or modernity. That is an old excuse for doing nothing; it is a myth, both cruel and inhuman."

It would be folly, Borlaug continued, to present millions of peasant farmers with mechanization when they had little understanding of machinery and lacked good soils and good plants. Most of them were so poor they took food from their mouths to buy a piece of cotton cloth for their backs. No, machinery was not the immediate answer. What was needed was a new spirit, a new crusade that would sweep the fields and convince governments that it was just as vital to support agriculture as industry or armaments.

The most precious commodity in Asia and Africa was human brainpower; to tap this power the FAO would have to inspire agricultural scientists to join the crusade to defeat hunger. New thinking, new materials, new men, new food for the land—these would offer the speediest solution.

"I propose the first order of priority should be the rapid, intensive, and practical training of a corps of dedicated young agronomists from each country willing to take part. The FAO can select the candidates and screen them for the best human material available. If there are countries that have no people with university training, then we should take the best possible and give them this training. I further recommend that, with the agreement of the Rockefeller Foundation and the Mexican government, these young men be sent, for roughly a year, to the establishments in Mexico where so many young agronomists have already been trained in the new aggressive approach to modern agronomy."

When implemented, the proposal produced Borlaug's "practical school for wheat apostles," which became part of the international wheat program. The FAO acted quickly and was winnowing out the candidates within weeks. The Rockefeller Foundation accepted responsibility for the costs, and the Mexican government agreed to make training facilities available.

The apostles were trained as were the early Mexican graduates. After passing through Borlaug's tough initiation process, they worked the first weeks with backs bent, toiling in the fields for twelve hours each day. Borlaug insisted, as he had with the agrónomos, that they work among the plants until, as he put it, the plants talked to them. They were given a sound grounding, by specialists in the growing Rockefeller team, in genetics, soils, agronomy, and breeding. They were taught how to level and lay out sample plots, where and how to sow the crops, and how to use fertilizer and water—how to be totally practical. Then they had to go back to their own lands and teach what they had learned.

The first wheat apostles came from countries in the Near and Middle East—Afghanistan, Cyprus, Egypt, Ethiopia, Iran, Iraq, Jordan, Libya, Pakistan, Syria, Saudi Arabia, Turkey—and from ten countries in South America. In the beginning the Indian government did not involve itself in the program. However, as the successes of the apostles in other lands began to make India's backwardness a conspicuous embarrassment, Indian scientists began to take the course in Mexico.

Field days had become fiestas in Mexico. Farmers, some from hundreds of miles away, arrived by the thousands to look at the displays at the research stations.

In the Yaqui each year CIANO was a venue that attracted streams of men—farmers wearing big sombreros banded with the colorful designs of their village. They came on buses and trucks, by cars and planes. In May 1960, with a fine winter crop of wheat harvested and hope of a bountiful summer ahead, the farmers flooded into Obregón and out along the long, flat road to the station between the Sierra Madre and the sea. Mexican scientists working under Narváez now handled the event. Careful planning went into the arrangements. Routes of inspection were laid out and marked with red, white, and green flags; flat-top trucks and trailers were rolled along the lines marked by the flags. At each stopping point an agrónomo explained the features of the plot display to the farmers as they stood on the flat decks of the trucks and trailers.

Borlaug flew to Obregón for the event soon after his return from Rome. He was on hand to see the dwarfs break out into the wheat lands of Mexico. He suspected what might happen when the farmers saw the experimental plots of semidwarfs. There had been heavy fertilizing of tall varieties in Sonora and Sinaloa and many fields of lodged wheats. The farmers' eyes would be keen to see the short stems. He told Narváez: "When they see those little wheats standing straight and proud, Nacho, you are going to have a real picnic on your hands."

Narváez said he knew the danger. The dwarfs were not yet ready and he did not want them scattered across Mexico's wheat farms. Dr. Alfredo García had been posted at the dwarf plots with several of the agrónomos to help him. The instruction was firm. Narváez said: "No one is to be allowed to get down from the flat-tops. They can look all they want. They can listen. But nobody is to touch." Borlaug did not want to interfere. He stood well back as the six flat-tops rolled up to the inspection point. García wanted to bring them all level with the plots, to give the farmers a chance to hear what he had to tell them. He called to the driver of the front truck to pull across and allow the driver behind to come abreast. He was planning to stand at the side of the flat-tops to tell the farmers to stay where they were, but he did not have time to give instructions. Even as he spoke to the front driver, between thirty and forty farmers ran down the path to the dwarf plots. From where he stood Borlaug could see their quick hands snatching at seed heads. Alfredo García and his agrónomos started to push the farmers away; the farmers pushed back. When García and his assistants tried to snatch heads of wheat from the farmers' hands, tempers broke and fists flew. García shouted for help. Farmers were thrusting their way out of the melee; some agrónomos fell among the wheats. Eventually order was restored; the sombreroed heads were reassembled on the flat-tops. The towing trucks moved away as García stood by the path, too angry to remember that he had not spoken his recitation on the attributes of the semidwarf plants.

By Borlaug's estimate a fifth as many heads of the dwarf wheats went rolling down the track in the farmers' pockets as were left standing in the experimental plots. He said to Narváez: "The dwarfs have escaped, Nacho. They are not ready yet, and you have a problem and a lesson to learn from this. You're going to have to move fast if you want to control this breakout, because those grains are not very good for breadmaking, and those farmers don't know that. You have to learn from this that you can never underestimate farmers. They're among the shrewdest people, and even when they can't read or write they're very quick to see possibilities."

The escape of this early type of Mexican semidwarf wheat could be countered only by swift action. "Put out your new, better varieties," Borlaug told Narváez, "the ones we are going to call Penjamo and Pitic. If they are better breadmakers than those that got away, the seeds you have lost will fall into a vacuum. The farmers will want the better ones; that way you get control again. But you have to work fast." Penjamo and Pitic still had defects that he had hoped to eradicate in the future breeding season. But there was no time for that now, and they went into production.

As Borlaug predicted, the farmers moved fast. When Narváez and Borlaug toured the main wheat areas the next season, they saw the wheats growing in clumps clear across the country. Many farmers had multiplied their stolen seeds during the summer and were increasing them again. Narváez was not ready to send out the Pitic and Penjamo varieties through the government seed agency; they would not be ready for at least another year. He faced his first major crisis as the leader of Mexico's wheat research program. He told Borlaug: "It would be wrong to let this wheat start appearing in the commercial bread grains without warning the millers. I have to tell them what to expect."

Late in 1960 Narváez and Borlaug called a meeting in Mexico City and explained the situation. Narváez recalled: "We told the millers' representatives what had happened, and that though we thought the semidwarfs should not be released because of the soft gluten and poor breadmaking qualities, we had to consider putting out these varieties. They probably would give trouble to the millers. The president of the association consulted with his colleagues. Then he came back to us and said, 'If these wheats will increase the Mexican harvest as much as you believe, then we feel it would be traitorous to Mexico not to release them. There are more than a million small families this would help. They are not organized like we are; we will get over the difficulty. We say—release the dwarf wheats!' We did. We wasted no more time in talk. We multiplied them and sent them out."

Narváez and his colleagues used every resource to obtain rapid multiplication of the two best breadmaking semidwarfs, Pitic and Penjamo, from Borlaug's breeding nursery at CIANO. Within two years, by spring harvest of 1963, the national crop yield was close to 2 million tons—six times greater than the yield in 1944. The semidwarfs filled

95 percent of Mexico's 1.5 million acres of wheat lands, and the yields on average-sized farms leaped to sixty and even a hundred bushels an acre.

Wheat productivity was to go higher still with better varieties—and with more intensive cultivation. Meanwhile, similar results were being reported by Orville Vogel and his colleagues in the state of Washington. By crossing the Norin dwarfs with local tall wheats, Vogel had produced a variety called Gaines, which later set a world record for wheat yield: 216 bushels on a one-acre trial plot. But though Gaines was highly successful in the United States and parts of Canada, it was a localized wheat and was photosensitive. Largely because the Mexican semidwarfs were photoinsensitive, they were to prove more suitable for cultivation over a greater area of the earth. But the Mexican strains still needed improvement.

Pitic and Penjamo had many of the qualities that Borlaug had been striving to breed into a semidwarf wheat. They matured early in the season. Because of that, and because of their photoinsensitivity, they could grow as far north as Sweden and as far south as Chile and virtually anywhere in between. They tillered strongly. They were resistant to the most ubiquitous races of rust. And they differed from earlier generations in that they produced edible bread: not good bread, not necessarily the most nutritious bread, but bread that could be eaten. Eliminating the soft texture and weak gluten of the grains, which were responsible for this poor breadmaking quality, was one of the major problems toward which Borlaug and Narváez directed their energies between 1960 and 1962.

But it was not the only problem. Short straws were an absolute requisite for a mass-produceable semidwarf wheat. In the dwarf seeds brought from Japan by Dr. Salmon the gene responsible for making short straws was recessive. It was linked to one of the genes that determined the quality of the wheat gluten, and the gluten, of course, was directly related to the breadmaking properties of the grain. In the thousands of crossings made in Mexico in the 1950s the project scientists had discovered that reducing the length of the straw through crossbreeding tended to make the recessive genes more widespread in the material and consequently more expressive. This was fine insofar as it affected the straw: it ensured an inheritance of short straws from one generation of wheat to the next. But it was not fine at all to the extent that it affected the second group of genes. For as the short straws became an inherited characteristic of the semidwarfs, so did the soft gluten. Moreover, in some breeds the ascendency of the two linked genes produced infertile offspring.

The manipulation of genes and chromosomes is an enormously complex, incredibly delicate process, but it was clear to the scientists that the two sets of genes had to be separated. Cleaving them apart had become, Borlaug believed, the final task that would turn the semidwarfs into the ideal wheat.

Between 1960 and 1962 Borlaug felt that they were on the brink of achievement. "There always seemed to be barriers beyond the barriers," he recalled later. "But I could not give up trying to get past them because I could see the light shining under the door leading into that genetic storeroom."

With success seemingly so close, the testing of every new generation of grains in the biochemical laboratory at Chapingo became a crucial process. Food values—the protein level and balance and the breadmaking quality—had to be established quickly in order to determine which strains of wheat should be planted in the coming season. Borlaug had worked out a chemical testing method with his cereal scientists that shortened the waiting time. One of the young

Mexican biochemists who cooperated in this timesaving advance was Dr. Evangelina Villegas. Others were Dr. Arnoldo Amaya and Dr. Frederico Chacón.

It was Dr. Villegas who led the team and who owed so much in her career to Borlaug's strength of mind and force of persuasion. In her student days, when she could speak no English and when domestic problems threatened to limit her efforts at study, he had convinced her to stick with it. He also made a personal effort to advance her career, as he did with many students. By the spring of 1962 she was the head of the new biological testing laboratories at Chapingo and was building a reputation as one of the foremost cereal biochemists in the Americas. Very attractive, in her early thirties, she was entirely dedicated to her work.

One day in May 1962 Dr. Villegas was standing outside her laboratory at Chapingo when she saw Borlaug striding quickly down the road. In his hand he held one of the familiar seed packets. There was nothing unusual about that: she had seen many hundreds of such packets; the filing cabinet in her office bulged with reports of analyses done on the countless lines of wheat seeds she had tested for him. But when he handed her this packet of seeds, his manner was different. She could sense his urgency and delight.

"Eva, aquí tenemos los mejores toros!"

Eva, here we have the best of the bulls!

Borlaug was not an easily satisfied man, and the words must have had portent for him to speak so. It had been a long, hard search. He told her these might be "the seeds of the future." It was a golden moment for Dr. Villegas; it meant much to her—much to her country.

The new seeds had already started a journey to Asia—so confident was Borlaug in his judgment. Ignacio Narváez had worked with him that spring in the Yaqui when the first class of wheat apostles was ending its year of training. The program loaded his hours with the extra work of teaching, demonstrating, and lecturing; and he still had to find time to initiate the international seed-testing scheme that was to grow into the International Spring Wheat Yield Nursery. Narváez remembered him in these days: "As well as training of the young men from different countries, he placed high importance on testing of seeds in the countries from which the apostles had come. With myself, and other colleagues, he sat for hours assembling collections of little packets of different seeds from the latest advanced varieties. These were for the apostles to take across the world with them to plant in the soils of their own lands. From that Dr. Borlaug would know how they responded to different environments."

Borlaug held a little ceremony at the Yaqui station before the apostles left for their homes. Each of them took a dozen or so seeds of the many new semidwarf strains. From that simple presentation the first seeds of change in Pakistan's grain revolution were carried to the experimental plots at Lyallpur, near the university, by Mansur Bajawa and Nur Chaudry. There they were to reveal the power of the genes that were at the heart of the Green Revolution. There they would provide the clue, ironically, for the agricultural revival in India.

The first semidwarf seeds arrived in India at about the same time, but by a different route. India had stood aloof from Borlaug's training scheme. Its first semidwarf seeds came for testing through the rust nursery system, which had been established on a widening international basis since the rust outbreak in 1948 in North America. The seeds went to the Indian Agricultural Research Institute in New Delhi, which the Rockefeller Foundation had helped to establish under Dr. Cummings.

There were about a dozen seeds of the semidwarfs, and they were divided between the IARI in New Delhi, the University of Agriculture at Ludhiana in the Punjab, and the Uttar Pradesh University at Pantnagar. Each institution was to grow them to test their resistance to local rust disease patterns. It was in the corner of the rust-testing plots of the IARI that the potential of the seeds was first understood. There were many "exotic" wheats—wheats from other countries growing there. Just the same, the four little bush-like wheat plants caught the eye of Dr. M. S. Swaminathan, then head of the institute's division of genetics. Later he was to direct the whole of India's agricultural research.

Dr. Swaminathan peered intently at the four dwarf wheat plants through his glasses. As was his habit when concentrating, he folded his arms and set his feet apart in a rock-like stance. Of medium height, with a broad forehead and a face of pale, quiet strength, he saw the four dwarf plants huddled among dozens of tall wheats from other parts of the world. They looked anything but impressive. They had not yet been given the supportive program of feeding and cultivation necessary to achieve full expression of their genetic power to grow more grain. But something about them intrigued him.

Borlaug and Swaminathan had followed similar lines of thought, on different sides of the globe. They had both faced rust as an enemy and had seen tall straws as a barrier to increased grain yield. Like Borlaug, Swaminathan saw in the little bush wheats a chance to turn hope into reality.

Borlaug said of him in later years: "He showed one of the most brilliantly swift agricultural minds I ever experienced. There was never any need to draw a picture for Swaminathan. His intellect and imagination put him ahead in his field; and not just in India, but in the world."

At the institute in New Delhi the Mexican semidwarfs showed the same resistance to rust; they grew without blemish, oblivious to their relocation halfway around the world. They responded vigorously to food and water. When they came to head, Swaminathan saw the set of their seeds and was at once busy arranging for the Indian government to issue an official invitation to Borlaug—through the Rockefeller Foundation—to make a tour of India's wheatlands and offer his advice.

Borlaug got the invitation early in 1963; within weeks he was flying east to the subcontinent. In mid-March Borlaug and Dr. Cummings left New Delhi to look at the main Indian wheat lands. It was swelteringly hot; the monsoon, less than two months away, was already bringing early humidity. The country was full of peasant farmers toiling to get the straggling winter wheat crops gathered in and the summer crops of rice, millet, or sorghum planted. The scientists went south toward Bombay, then northward into the Gangetic plains to the Punjab. They saw the great Indus water scheme over which India and Pakistan had wrangled three years earlier. They saw some of the long reach of the Ganges, which runs for 1,000 miles through the most heavily populated area on the planet. They caught sight of the great Brahmaputra River to the east and came back through Bihar and Uttar Pradesh. And everywhere they saw the ache for rain. All depended on the monsoon; it was the arbiter between food and starvation.

It was five weeks of arduous travel. Borlaug said later: "I came out of that tour certain that the little people could never do it on their own, no matter what materials we gave them. There were much bigger, much greater factors to be settled at the top level before we'd ever have a chance to get the wheat revolution off the ground."

He returned to New Delhi late in April and consulted with Dr. Cummings and Dr. Ernest Sprague, who had taken over the post of joint-coordinator of the Indian corn improvement program from U. J. Grant in 1959. After a day or two Swaminathan asked him to join a small conference he had arranged. Dr. B. P. Pal, Dr. A. B. Joshi, Dr. Cummings, and a number of people from the institute were there. They listened to Borlaug's comments and discussed the relevance he saw between India and the Mexican experience. Dr. Joshi eagerly pressed the analogy further. As recently appointed coordinator of the Indian wheat improvement program, he was forming his plans for the introduction of a countrywide series of wheat trials, which were to be run on a three-year basis.

It was Swaminathan who came first to the question of the dwarfs. "Now that you have seen much of Indian wheat, do you think your new dwarfs would fit this land? Could they succeed here, as in Mexico?" he asked Borlaug.

Borlaug sat in the room in New Delhi as the others plied their questions. They were asking him to declare with certainty that the dwarfs would be a central factor around which they could transplant the whole Mexican triumph into India. He could not give them this assurance. Caution was necessary.

"There are imponderables," he told them. "We don't know much about your soils. Too much information is missing." There were other considerations. The disease resistance spectrum in the dwarf wheats had been bred carefully to prevent disaster from rust pathogens in Mexico. How the wheats would behave in India's humid heat with other races of rust was uncertain. And there was the price of grain. The farmer must be given rock-fast guarantees that when he produced more he would be paid more, and not be sucked dry by the grain market manipulators. And how would the new varieties perform with the primitive Indian approach to cultivation?

They were all looking at him, waiting when he thought how he might find the answers. But he was in India, and he could not say what was on his mind. He had thought of the two fine young Pakistani apostles, Nur Chaudry and Mansur Bajawa, and of another apostle who was growing the dwarfs under similar conditions, the first-year graduate from Egypt, El Ham Talaat, who was working at a station north of Cairo.

He smiled directly at Swaminathan. "I think there is a way I can find out some of the things we need to know very quickly. Give me a few weeks and I will write you a report and set out my opinions. I am sorry to delay, but I must be able to answer your questions with more certainty than I can at this time."

He flew from New Delhi to Karachi. After making inquiries on the whereabouts of the two Pakistani trainees, he took a flight to the town of Lyallpur, the site of the Agricultural Research Institute and the university where Bajawa and Chaudry were working. He was met there as an honored guest, not by the wheat apostles but by the senior scientific administrators of the research institute, who knew his reputation. He wasted no time and asked to be shown how the Mexican wheats were performing. They led him, followed by the two apostles, out to the trim, level plots, where he found himself looking at a small row of his dwarf wheats.

They were a total disappointment, limp and drooping, poorly tillered, totally unimpressive.

He knew his wheats well. They were not yet perfect, but they were infinitely better than these unhappy, struggling performers in front of him. He turned to the Pakistanis, his astonishment clouding his tact for a moment, and blurted out: "What in hell is *this?* You haven't fed these damn things. They haven't been properly fertilized!"

One of the administrators told him, respectfully, that all their experimental plots had a ceiling limit of fertilizer. "We never use more than the ratio of thirty-five pounds of nitrogen per acre on this establishment," he said. "That is a firm rule."

The young scientists remained silent. An air about them,

something in their eyes and their faces, told him the story. It was the recurring problem of seniority as a brake on progress. He said nothing more and walked to the next plot of wheats. He was given no chance to talk privately with the two young Pakistani scientists. He was hustled away to dinner and entertained into the evening until he pleaded tiredness and was shown to his room in the research institute staff house.

A tapping on the window, quiet but insistent, brought him out of his sleep. He climbed out of bed and pulled back the curtain. The first lick of day was across the hills, enough light for him to see Chaudry and Bajawa standing there, fingers to their lips for silence. They motioned to him to dress and follow them. Quickly he pulled on trousers, shirt, and boots. The three of them made their way across the lawn and out into the experimental plots. "Single file, without a word," he recalled, "they led me across the fields to a far corner. The sun was not yet up, but there was light enough for me to see that small green patch. There they were—the dwarfs—thrusting up their heavy heads of grain, jostling and moving proudly in the slight breeze, happy and vigorous, and as beautiful as I had ever seen them growing in the Yaqui. Unknown to those stuffy officials, these two courageous young scientists had sown a second generation from their samples and had fed them the correct amount of fertilizer. The plants looked beautiful. The two young men had had to complete the masquerade. They called them by another name, a name not even hinting that they were of Mexican origin. I could have danced! We shook hands; we laughed together. I took a closer look at that little green patch and knew in a moment we had something on our hands for Pakistan—and for India. A little green patch in a corner—all unofficial. It verified what was possible, provided everything else came out right."

A moment of joy for him, a vindication of the delicate

task of breeding wide adaptability into the dwarfs from Mexico. Here, on the other side of the earth, they grew quite happily. He needed little more confirmation; he sought it anyway.

The next morning he flew to Cairo and found the young Egyptian on a station about sixty miles north of the city. The dwarfs were doing well in a small plot. But here again senior scientists retarded their full growth by permitting only a third the needed amount of fertilizer. The young Egyptian's hands were tied more tightly than those of his Pakistani colleagues. They had not been untied by subterfuge.

Borlaug flew at once from Cairo to Rome and rested for two days, after making a courtesy call at the FAO headquarters. He consciously struggled with all the possible complications. A massive transplant of plant materials, technology, and methods across the world had never been attempted on the scale he contemplated. The disease problem had to be taken into consideration. Were there unusual rust pathogens in other parts of Egypt and the Indian subcontinent that would soon penetrate his Pitic and Penjamo strains—and the ones coming behind? In his head he traced back the long lines of resistance he had bred into his races. It was possible —not probable—that unknown enemies lurked. Yet he had to be bold to succeed. What would it mean if India, like Mexico, could grow seven times as much wheat in the next ten years? That would be a victory against famine!

Heading back to New York, he sat awake all night as the aircraft droned high over the Atlantic. He was thinking, and his thoughts flew fast. Hundreds of those mighty little dwarfs, growing happily in the morning sun in Lyallpur. And he was saying to himself, over and over again: "This thing can go! It can really take off!"

Twelve

His mind racing, his spirits soaring, Norman Borlaug returned to the Yaqui Valley to talk to his plants. It was April 1963, and the last days of the plant-breeding season were melting in the warmth of an early summer. The sky was cloudless and very blue; the wheat fields glared in the bright sun. His mind was on India, but what would happen in India depended on what was happening here.

Three men crouched in the distance, their heads bent over plants. Borlaug realized that he probably wouldn't know them; a new group of wheat apostles had come while he'd been gone. He approached them anyway. "I'm Borlaug," he said. "How're things going here?" The trio rose respectfully to their feet. One of them extended his hand, a wide grin splitting his dark chocolate face. He was from Kenya, he said, and his two colleagues were from Afghanistan and Turkey. As they spoke about their work, they were shy, almost reverential. They were in awe of him. But he didn't want to intimidate them. He must retain their confidence; he must see that they were comfortable and secure around him, for they must never hesitate to speak frankly to him, to ask him questions, or to reveal their ideas. And so he was very friendly to the three agricultural apostles—friendlier probably than he would have been a few years earlier, when he was less famous.

The work in the fields, Borlaug was satisfied to learn, was virtually unceasing. Thousands of strains of wheat had been crossed, and the offspring were growing under the care of the apostles on the experimental plots. All day long the fields echoed with the sounds of countless tongues, and the multicolored work force was wisely supervised by a master wheat technician, a young Mexican—none other than Reyes Vega, the eldest of the three little brothers who had been hired years before as bird boys.

Important visitors from many countries appeared regularly to see the now famous wheat station. Narváez was glad to show them around, but they made his life more difficult. His days had not been long enough since the big breakthrough of the previous spring—the cleaving of linked genes in the semidwarfs the previous year had been the major development in the effort to make a wheat with better breadmaking capability. Now his days were shorter still, with tours to conduct and hundreds of questions to answer.

Borlaug and Narváez were trying vigorously to further perfect the new strains, which Borlaug had named Sonora 64 and Lerma Rojo 64. One improvement they were seeking was an even shorter straw. Shorter straws, claimed Borlaug, would inspire more extensive tillering. Furthermore, plants with shorter straws would mature earlier, harvest more easily, and mean less work in winnowing and threshing. Most important, if additional fertilizer were fed to the shorter plants, the dwarfs would offer a greater return in grain.

Now, Borlaug's goal was to breed a version of the semidwarfs that had a shorter straw, kept its good breadmaking quality, and remained fertile. Borlaug had known for some time that shorter straws could be achieved by joining the gene that insured short straws with a second or third similar gene. The ideal would be a two- or three-gene dwarf wheat that would respond to extremely heavy feeding of nitrogen fertilizer. Previous experiments had produced wheats with astonishingly high yields; but they had also produced many sterile offspring. He would have to cross many lines to achieve it, but that was what the tens of thousands of rows of plants were there for—to experiment with. So all day long he walked among the rows, listening to the plants, finding out what he could from them. In the evening he pored over all the records and data about the plants, and late at night or early in the morning he worked on his report to Dr. Swaminathan. One man in Mexico, he was trying to make food for 500 million people across the world. While Borlaug was gathering his thoughts for the report he would make to Swaminathan, another Rockefeller report was being written in New York. Harrar was the new president of the foundation and his hand was on the report being prepared for issue by the foundation trustees. It read:

"About half of the human beings on earth have an inadequate diet, and millions live constantly on the edge of starvation, despite the fact that an overabundance of food is produced in a few technologically advanced countries. A world that possesses the knowledge and methods to confront the demands of hunger must accelerate its efforts to increase the production and improve the distribution of food.

"We propose during the years ahead to strengthen that part of the foundation's program that is directed to human nutrition. This will mean increasing the quality and the quantity of those foods that feed the world. . . . It will involve the search for new knowledge leading to better use of land and water resources, development of nonconventional agricultural techniques, and investigation of the changing environment in which we live."

The Rockefeller-sponsored campaigns against world hunger had expanded dramatically in the two decades since Harrar had arrived in Mexico to open the foundation's first rural improvement program. Mexico had been the launching pad. The work in wheat and corn was extended to Colombia, and to other South American and Central American countries. Scientists—Joe Rupert among them—followed the patterns developed in Mexico to bring new plants and new methods to Chile, Peru, Honduras, Nicaragua, Costa Rica, El Salvador, and Panama; and this was only the beginning of wider operations across the world. Wellhausen, Harrar, Bradfield, Stakman, and many others had taken expertise and financial help to countries in the Middle East and Asia. Collections of plant materials, wheat and corn, had also been sent to breeders in Kenya, Ethiopia, Poland, Indonesia, and—among a dozen other countries—India.

In the late 1950s Harrar, Weaver, Mangelsdorf, Wellhausen, and Robert F. Chandler—the young scientist who had worked in the early Mexican program—went touring the hunger spots of Asia to evaluate the enormous rice problem. Their findings had led to the establishment of the International Rice Research Institute at Los Banos, in the Philippines. It had Rockefeller support and backing from the Ford Foundation and the Philippines government; but the basic approach to the problem was based on the experience of the original team in Mexico. Robert F. Chandler, the institute's first director of research, adopted lines of attack that Borlaug had developed for wheat in the Yaqui. The search started for short-stem rice, in which higher yield and greater resistance to disease could be bred.

Rice is by far the most important single food source in Asia. Effective research into this crucial cereal required extensive investment of time, money, and scientists. The two American foundations carried the burden; the government in the host country provided the site and certain services. In its effect on world food problems the work on rice became a new and powerful part of the burgeoning Green Revolution.

At that time the average Asian ate more than 300 pounds of rice each year. (In the West the average annual consumption per person was six pounds.) So prominent was the role of rice in daily life that the rice research at the institute could not help but have political significance. As Borlaug was to constantly hammer in his international statements, grain foods and peace could never be separated in world affairs. The demand for rice was so great, so continuous, that there were rarely any carryovers from year to year; rice was highly vulnerable to drought, to damage from weather, and to certain fungi, viruses, and insects. The work of the institute was a vital contribution to a new age in food production ability.

Borlaug was staying in a hotel in Ciudad Obregón—not a good hotel, but he could think and write there. He was in his fiftieth year, still hard and fit. Yet bending and sweating along the rows, being the first in the fields and the last out, and setting an example to men half his age was physically trying. He went alone to his room each night for the comfort of a hot bath—and the writing of the Indian report. Swaminathan was waiting.

He thought about the eight to ten bushels an acre, on average, that the tall Indian wheats were yielding—while Mexicans were throwing more than eighty pounds of nutrient fertilizer an acre on their lands and registering yields of eighty to one hundred bushels. Indian farmers were still tilling with wooden plow and buffalo, and they were too poor to buy fertilizer. Would it be sensible to suggest a massive transplant of Mexican methods and materials across the globe?

Faith, confidence, and boldness were needed. He started

his report to Swaminathan, thinking that it was certain to go through Cummings' hands for polish, for toning down. (Actually, Cummings sent it as it was: blunt and to the point.)

"The evidence at this time is circumstantial; but fragmentary as it is, there is the prospect of a spectacular breakthrough in grain production for India. The new Mexican varieties that I have seen growing incline me to say that they will do well in India and grow beautifully. But though such opportunities do exist, there are conditions that must be met with fact and action—and not with talk and probability. If the disease resistance holds up in its present spectrum—and we can check that quickly—and if the quantities of fertilizer and other vital necessities are made available, then there is a chance that something big could happen.

"I must warn, however, that senior scientists and administrators must not hold this thing in check and must not restrict the ardor of the bright young scientists. If they do this, they will cause roadblocks and bottlenecks for which the poor will pay the price in continued hunger."

He sent another letter, similar in content, to the people he had met at the Lyallpur research center in Pakistan. He told them that India was about to make a decision on a new injection of material and money into wheat. He used that tactic from the start, counting on the fierce, antagonistic rivalry between the two nations to stimulate food production. His justification for doing so was simple: no one could build a peace on empty bellies.

He followed the letters with practical action. Scouring his fields and seed collections for his most rust-defiant wheats, for the best-yielding varieties, and for others he thought might do well in their new home, he chose the latest line of Lerma Rojo—not then released commercially in Mexico—Sonora 64, and several others. He sent a variety of smaller samples of more advanced lines. In all, he dispatched 900 pounds of experimental seed to India, 450 to Pakistan. He sent them by air to be certain they would be there in time to be planted for the coming fall tests. Time was running out.

He was not alone in his sense of urgency. Many experts forecast doom for millions in massive famines that would sweep the world in the mid-1970s. While these fears were appearing in print, the seeds of change had already been sown and were being carefully fostered in Asian soils.

The wheat research program in Mexico was altered again late in the summer of 1963, when President Adolfo López Mateos returned home from a global tour. In most of the agrarian nations he had visited Mateos had received accolades for the advances that had been achieved in his country. He heard and saw firsthand how methods pioneered in Mexico were at the heart of a powerful and creative movement in world agriculture. Back in Mexico City Mateos ordered a state dinner at which he thanked Harrar, Wellhausen, and Borlaug for their devoted work and expressed his pride at what he had seen abroad. He now felt, he said, that the work being done under the Inter-American Food Crop Improvement Program—which had more or less replaced the work of the old Rockefeller project—should be expanded, and he wanted to encourage the formation of a new organization whose work would have a greater international character. The organization—the International Center for Maize and Wheat Improvement—was established quickly: on October 25, 1963, Mexico's minister of agriculture, Julian Rodríguez Adame, and Harrar signed the agreement formalizing the cooperative enterprise.

Based at Chapingo and in the Yaqui at CIANO, its work attracted wide support from a growing number of philanthropic and government agencies. The Ford Foundation added its weight; so did the U.S. Agency for International Development and the Inter-American Bank.

Out of the wheat center sprang CIMMYT, the acronym for the Spanish title Centro Internacional de Mejoramiento de Maíz y Trigo—the International Corn and Wheat Improvement Center. Wellhausen was its founder and general director, Borlaug the director of its world wheat program. The board of trustees included prominent figures from many lands, and the name CIMMYT became synonymous with efforts to benefit all the peoples of the world.

A nonprofit organization, CIMMYT was responsible for training personnel and introducing the benefits of research on a global scale. It widened its teaching and its services to wheat growers across the world. It turned its offices, laboratories, and test plots into clearinghouses for new crop varieties sent from more than a hundred nations. It made all its data and materials available free of charge to every country on earth.

CIMMYT came into being in the mid-1960s with a multimillion-dollar budget and extensive new laboratories and buildings—a complex now called El Batán. It grew up close to the old adobe shed with the tar-paper roof where Borlaug had slept in 1945 amid the bags of decayed vegetables and scampering rats.

The years had brought changes to Borlaug's home too. He was an international traveler now, and his long absences occurred just at the time that the children were leaving the nest. Margaret Borlaug's life had been lonely ever since they'd moved away from Wilmington. Now she was to be completely alone for months at a time.

Jeanie had become a beautiful, intelligent, spunky young woman. She adored her parents. As soon as she had been able to understand her father's work, she had become a vocal champion of his causes, and she had always been extremely close to Margaret. Full of vitality, surrounded by friends in the American and Mexican communities, Jeanie seemed able to take advantage of every opportunity that life offered her. She was easygoing, independent, and devoted to her brother. When she left for the University of Kansas—where she majored in education in order to become a high school Spanish teacher—Margaret was sad to see her go but confident that Jeanie would manage to make the most of the new stage of life she was entering. Margaret had never worried about Jeanie the way she had about Billy.

Yet even the worrying about Billy had decreased. His childhood allergies and overall poor health seemed to have passed. There was nothing delicate about him now; he had suddenly sprouted upward and outward. He was six-foot three, weighed 200 pounds, and was strong and sturdy. But he was more introspective, less life-embracing than Jeanie and had not enjoyed his adolescence in Mexico. A year before Jeanie's departure, in 1962, Billy had left home and gone to a boarding school in the United States.

The departure of the children coincided with the increased demands on Borlaug. There were long days and silent nights for Margaret. She revived her appetite for reading, racing through book after book on every conceivable subject.

She said much later: "It got so compulsive in time that I even started reading the small print on the cereal boxes at breakfast."

Reading was not enough. Now that they had more money —Borlaug's salary was raised to the level of a professor in the United States—she could afford occasionally to fly north to visit her brothers, her father, and Norman's parents and sisters in Iowa. Her own mother had died long before. She also took tours to historic sites around Mexico City.

Margaret came to terms with her new emptiness. This time in her life required more adjustment than any other, but she later recalled: "I had to take the good with the bad, being married to a great man—though perhaps I say that out of blind love. I knew he had made the breakthrough with the dwarf wheats, that he had started the school for wheat apostles, that he had charge of the biggest experimental wheat station in the world, and that everyone was yelling for some of his time. I am not a scientifically minded person, but I saw what was happening to him. He was away for periods up to three months, and I knew that wasn't going to get less as time went on. I knew he needed a home, a place to come to where he would always be comfortable and happy. I tried to make it like that, not to make demands on him or his time. I suppose if I had those years over again I would want to make more demands on his time—but I wonder if I really would. I had to make my contribution. I hope it helped."

Late in 1963, then, she was alone with all the memories of the children's days, while India and Pakistan were calling to her husband for more help.

In the fall of 1963 Borlaug's consignment of experimental semidwarf seeds were sown into the brown soil of India. The initial planting occupied just twelve acres. But they grew so vigorously that Dr. Swaminathan was able to calculate, from his plots near New Delhi, that the yield would be ten times the conventional wheat yield per acre. And in Pakistan, near Lyallpur, Bajawa and Chaudry continued to feed their five-acre plot as they had been taught in Mexico. When they saw the results, they aggressively pursued the senior scientists at the Lyallpur Research Institute to show them what was happening and to convince them that they were on the threshold of a new era in food production.

The conservative Indian scientists had fed their different plots with different levels of nitrogen fertilizer. But in the spring of 1964, when they compared the dwarfs with the best of their tall varieties of wheats, the superiority of the Mexican strains—even those that had been undernourished—was conspicuous. The genes of the dwarfs had returned to Asia with unprecedented growth and endurance. Dr. Swaminathan began pressing the government to initiate a wheat development agreement with the Rockefeller Foundation. When the government responded and sought a wheat program similar to Dr. Cummings' corn improvement program, the foundation asked Borlaug to fly to India to survey the possibilities.

In New Delhi Borlaug met with Dr. Pal, Dr. Cummings, Dr. Swaminathan, and Dr. S. P. Kholi, who had been named as Indian coordinator of the wheat improvement program. Harmony, good humor, mutuality of interests, and most of all optimism characterized the meeting. According to the recollection of Dr. Cummings, the most realistic and sober man at the meeting was Borlaug. "Seeds are not miracles," Borlaug reminded the scientists in the room. "Ideas are far more powerful. The seeds are the catalyst. You have to support them with a total package of modern technology—and with strategy, with psychology, and with firm, guaranteed economic policy from the government. The farmer has to *know* what you're doing. You can't play fast and loose with his faith. In the end he's the guy who has to grow the food. He has to get the proper reward for his effort."

Aside from this warning, the meeting was a quiet, comfortable start to the new project.

In the spring of 1964 the Green Revolution was beginning in India, and everyone involved knew it. Borlaug found that the most powerful politicians in India understood it: they welcomed it. The enemy, as predicted, was the government scientist, the tory who feared the future and the possible loss of his power.

One day a senior scientist, a man who had climbed the

bureaucratic ladder to one of the top positions in scientific administration, led Borlaug into his office. He looked solemnly at the American and came to the point: "I suppose you are sincere—I will assume so, anyway, out of respect. But just suppose there is something that goes wrong, some unseen thing in the changes you cause? Suppose some imponderable leads to disaster? Where will you be then? You will not be here when that happens. But we will! We will have to face the penalty—we, and our families, who depend on us for their security."

That same night Cummings held a gathering of important people. One senior government man said he thought Borlaug would have trouble getting acquainted with India's customs and need. Borlaug was blunt: he said that he understood all human bellies to be constructed on the same blueprint. He had no regret that he was brusque.

Borlaug and his colleagues had to defend their actions because their operations challenged the traditional approach. The Mexican seeds were the catalyst. But fertilizer was the fuel to give the grain its thrust and power. Yet the use of fertilizer was vigorously opposed. Most vehement of all in the struggle were the academic and agricultural economists. The stage was set for a titanic battle of wills. Borlaug told Cummings that they were on the side of the big battalions. There were hundreds of thousands of farming people in their ranks. He also identified some of his main critics as "experts" who argued that Indian farmers were so ultraconservative that they would never be moved from traditional farming—another facet of India's fatalism.

Borlaug flew directly to Lyallpur. The atmosphere there was alive, different from that in India. People seemed keener, more vital. They seemed harder and more disciplined than the Indians. But the edge of hunger in India was sharper, more cutting than in this land, and that does something to toughness and ambition. At Lyallpur he had a receptive audience. This made it easier to promote ideas and to argue that it was the time for new technology.

He went out with Bajawa and Chaudry to see the semi-dwarf plots. He was so pleased that he stayed on to give a brief seminar and to walk with a party of young agronomists through the extensive plots of different wheats brought in from across the world. He was so entranced that he forgot it was his fiftieth birthday. But he received an important gift: a colleague and a powerful ally in Pakistan. On the edge of the group, listening as Borlaug talked, was a square-built European with a serious, strong face; his thick black brows were drawn together as he scribbled notes with a chewed-down stub of pencil. He was at the back of the group when they stopped by the best display of Sonora semi-dwarfs, strong with swelling grain heads. Borlaug swept his arm over the little sea of ripening wheats and said:

"This is not a stroke of luck! There are twenty years of aggressive research behind these wheats—research that enabled Mexico to go from hunger to self-sufficiency, to export grain foods. And what Mexico has done, you can most certainly do in your country—only you can do it much faster!"

He hammered home his point. There were no miracle seeds. Nothing the generations of wheat breeders had done, nothing he had done, had created a single new gene. Human skill had been managing, manipulating the basic material that had been at hand for all of history.

As Borlaug walked through the little throng, the man taking notes pushed forward and shook his hand. He identified himself as Haldore Hanson, senior representative of the Ford Foundation in Pakistan.

They had their evening meal together in the same dining room where, one year before, Borlaug had carefully avoided the controversial issue of the fertilizer ceilings imposed by rigid minds. Now there was no such restriction. He talked without inhibition on the genetic potential bred into the dwarfs. He said they would repay handsomely for the food applied to their soil. He had double-gene dwarfs—and possibly triple-gene dwarfs—on the way, that would respond to still higher dosages of fertilizer, without the penalty of lodging, which was common with the tall wheats. There were two ways to look at the result. Each ton of fertilizer used on the wheats, along with the necessary irrigation, would return between twenty and thirty tons of *additional* grain. Each ton of fertilizer used on the wheats would also drastically reduce the area of land needed to feed the people.

At the end of the evening Hanson invited Borlaug to his room for a nightcap. As Borlaug remembered it, Hanson took a bottle of Scotch from his briefcase, poured a liberal amount in two glasses, handed one to Borlaug, and said: "Did I hear you right out there? Were you saying a revolution in food production is possible in this country—or was I mistaken?"

Borlaug looked at him for a moment over his glass.

"You heard correctly," he said. "The fruit is only waiting to be picked."

Hanson pulled up a chair and took a sip of his drink.

"I think we might have some talking to do," he said.

They talked on until well after midnight. The Ford Foundation had no agricultural program in Pakistan. It was working in support of education and family planning against the population explosion. But here was Hanson keen and eager to propose a wheat program to the Ford Foundation. Borlaug could participate in the program as an adviser through CIMMYT and Rockefeller—if that was acceptable to the Pakistani government.

Hanson knew many of the people who would be involved in a decision to enter into a joint wheat program. He was a former newspaperman with wide experience in the Far

East. He showed his professional flair and skill in compressing Borlaug's data into factual, brief, hard-hitting points. He wrote his report and on the following day contacted Pakistan's minister of agriculture, Malik Khuda Bakhsh. The Pakistani was intrigued and invited Hanson and Borlaug to his home in Lahore for tea and an afternoon discussion. That afternoon Bakhsh's home was filled with Pakistan's agricultural brains—agronomists and botanists from universities and government departments, experts in extension services, and administrative scientists.

The audience was skeptical and largely self-interested. Borlaug was heard in silence. The atmosphere was heavy with rejection, although the antagonism was unspoken. Borlaug and Hanson left knowing that opposition would come into the open once they had gone, that wrangling over their proposal would be bitter. They suspected the weight of argument would go against introduction of the dwarfs and the new technology. (Malik Khuda Bakhsh confirmed this, much later.) Left to the advisers, the wheat project would sink without trace. But Malik Khuda Bakhsh was a cautious man. He argued that it was out of character for men from two great American philanthropic foundations to make a proposal that had no basis in fact. Also, he knew that across the border in India there was interest in the new wheats. It would be wise to keep the options open. It was fortunate for him, and for Pakistan, that he did. Soon afterward President Ayub Khan read a newspaper report dealing with the promise of new Mexican wheats in India. When he asked why Pakistan had not tried the new wheats, the secretary told him that negotiations were already under way with the Ford Foundation. On such thin threads hung the Green Revolution.

After the meeting in Lahore Borlaug and Hanson realized that implementation of the project in Pakistan would have to rest on a man of special skill and missionary zeal—a man who had the special qualities of patience, drive, and leadership and who was experienced in Mexican wheat technology. The same was true for the project in India. "What we need is a couple of tough, determined, scrambling quarterbacks," Borlaug said.

They would have to be top-flight men, courageous men. Men who would take the blows, the veiled disbelief, and the objections and still go on; men who would work under pressure in the fields when the sun was blazing and their bodies were running with sweat, and when the monsoon rains were smashing into their faces. And they would have to do it a long way from home.

At that time Pakistan was importing more than a million tons of wheat each year—a great portion of which was gift wheat from the United States. But that did not make Americans popular. Hanson was convinced that a non-American agronomist should head the project for Ford in Pakistan. But where would the Ford Foundation find such a man? Borlaug said he thought he knew the best possible man for the task. Even at that time this man was packing his bag in Mexico for a tour of countries that had sent apostles to the school in the Yaqui. He would soon be in Lyallpur. "If CIMMYT agrees, and Mexico agrees, and Rockefeller and Ford agree, I'd suggest Dr. Ignacio Narváez for the job—and you couldn't get a better man."

There were many weeks of negotiation, preparation, and paper work before Hanson could get the path cleared to set the Ford wheat project into motion. In the meantime, he met Dr. Narváez and fully concurred with Borlaug's appraisal of the proposed field commander for Pakistan.

Flying back to the United States from Lyallpur, Borlaug worried about the field commander for India. Planting the superwheats across India would require a man of superlative character and training, a cool man with courage and the heart to fight. He could think of no one better qualified than Dr. Glenn Anderson of the Canadian department of agriculture.

Borlaug had met Anderson at the first world genetic symposium in Winnipeg in 1958. Anderson had since been to the Yaqui several times, working on materials in the Canadian rust research exchange program, and Borlaug had noticed him working in the fields when others had caught the plane back to Mexico City. He had tested Anderson's mettle by devil's advocate arguments, and had once taken him fishing along the irrigation canal, seeking out the strength of his convictions. Anderson filled the bill. But was he available—and willing?

The Canadian scientist was ill with hepatitis when Borlaug called his home. Anderson was very interested, and before he was completely well he flew to New Delhi and met Narváez, who was in India as part of his tour. Anderson immediately liked Swaminathan and went with him on a brief tour of the northern plains and valleys along the Himalayan foothills. Late in April he flew back to New York, telephoned his wife, and asked her to start packing.

He and his family reached New Delhi on August 15, 1964, in the middle of the monsoon season, when ten inches of rain fell in a few hours. There were torrents in the streets, but the entire Rockefeller staff turned out at the airport to greet them. Dr. Cummings was there too. He said to Glenn Anderson: "Borlaug hired you because you know your business. India does not have enough to eat. You are to remedy that situation. How you achieve it is your decision."

Three days after arriving in the steaming city Anderson attended a meeting to discuss the Indian wheat improvement program. He and his Indian coordinator at once proposed a new approach to national crop yield development. They divided the country into five zones of action and laid the basis for technical, financial, and administrative implementation of Borlaug's plan. The program would have to be a tough one. The experimental plots would have to be set out on hundreds and hundreds of acres so that they could be seen by farmers across the country.

In India Anderson was at work, and in Pakistan Narváez had started operations. Meanwhile, back in Mexico, Borlaug had brought the pieces together in the new strains of semidwarfs.

By the summer of 1964 Borlaug's pessimistic view that the hungry nations might never solve their food production problems was dissipated. Seventeen acres of green dwarfs growing on the Indian subcontinent had routed his doubts. In August he was at Purdue University in Indiana to deliver a lecture at a historic meeting of the American Phytopathology Society. Biochemists at Purdue had made a startling advance in the development of nutritional grain. Now Borlaug was predicting that food output in India and Pakistan could be at least quadrupled.

The Purdue discovery was a breakthrough of the classical type in research—a chance observation meeting—a prepared mind. Dr. Edwin Mertz of the university's biochemistry department had instructed one of his doctoral students, Lynn S. Bates, to make a chemical analysis of the amino acid content of certain corn grains. Among them was a mutated corn type, Opaque 2, which had long been known as a curiosity. It produced soft, fluffy corn flour and was not very popular with consumers. Bates did the work and reported, with surprise, that Opaque 2 had given a reading for lysine that was double the amount in normal corn. Mertz told him to make the test again—properly. Bates came back with the same result.

In a flurry of excitement Mertz realized that the consequences could be immense for a wide segment of humanity—the millions of people who subsisted on a corn-based diet. It was a crucial observation, because lysine is a key substance used by the body in building up protein molecules from amino acids. Biochemists have produced the "shortest stave in the barrel" analogy to explain how amino acids such as lysine fulfill their role in nourishment and in the repair of living tissue. More than twenty amino acids go into the protein assembly, but about half a dozen are more crucial than the others. Of these lysine is one of the most important, and one of the most common. Because the production of protein is limited to the amount of lysine available, no matter how much of the other amino acids is provided, the protein barrel cannot contain more than the limit set by lysine—the shortest stave. Thus doubling the lysine content of a kernel of corn was tantamount to doubling the value of its protein.

So epochal was the breakthrough that the Rockefeller Foundation immediately started a project for testing Opaque 2 on undernourished animals. Later this work was extended to reviving half-starved children in the Caúca Valley of Colombia. And in Mexico Borlaug's associates opened a program to improve the dietary appeal of Opaque 2.

Before Borlaug delivered his lecture at Purdue, the news of the protein discovery had come as a stimulating surge of hope. India faced the most critical food problem on the globe, with more than 460 million people existing on 385 million acres of arable land that gave the lowest average grain yields of any country. As things stood in the underdeveloped world, three out of five human beings went to bed hungry each night. The massive calorie deficit was equal to the entire annual grain crop of the United States plus two-thirds of the Canadian harvest. But that was not all. Lack of protein damaged the nervous system. The damage stayed for life. It afflicted millions—even those who appeared to have sufficient grain to stay alive. It was deadly in the young; it slowed the fully grown. It could be, and often was, lethal in infants. The nervous system develops 80 percent of its potential before the age of eighteen months, and the brain achieves 90 percent of its growth by the age of four. The vicious fact was that once children had suffered lack of protein during infancy, the damage was irreversible. It handicapped them for life.

Borlaug had seen the signs in many children—swollen bellies, bloated legs, black hair tinting to red, and watery, lackluster eyes. Of children suffering protein deficiency, one in ten or twelve died in the first year of life; many more died before they reached the age of four. The children did not receive the protein they needed to repair and build their bodies. A diet of only grain could not provide the balance needed for protein nutrition, since the key amino acids, lysine and tryptophane, were in short supply.

Thus the news of the Opaque 2 lysine discovery presented another challenge to Borlaug. Could they not find higher protein levels in other cereals? The breakthrough had come from a mutation in hybrid corn that provided the building blocks for protein. Would this be possible in other cereals?

His mind was racing all through the Purdue meeting. His thoughts went back to the Yaqui Valley, to the day in 1945 when he had sat by the irrigation weir and pondered crossing wheat with other cereals. Then he recalled some strange tall plants he'd seen by chance in a laboratory in Winnipeg, at the University of Manitoba, back in 1958. Hybrids of wheat and rye, they were called triticale, after the Latin *triticum* (for wheat) and *secale* (for rye). There was a gulf between the chromosomes and genes of the wheat and rye plants, but they could be induced to come together with artificial crossing. The problem was that the resultant hybrid plants were always infertile. But by treatment of the growing point of the young hybrid seedling with colchicine, a chemical extract from a common type of spring crocus, partially fertile plants resulted. Many geneticists said it was a bridge that could never be crossed with any hope of developing a commercial crop, however. Since nature had not overcome the partial sterility barrier in tens of thousands of years by natural pollination, there seemed little hope of man finding a way. But Borlaug thought he should try.

Triticale had been known since the late nineteenth century, and work had been done on them by Muntzing in Sweden, Sánchez-Monge in Spain, and Omara in the United States; Winnipeg breeders had been interested for some years. The plants were a curiosity, an oddity. Now, inspired by the corn protein breakthrough, Borlaug's imagination saw promise of potential. Rye was a plant that grew well on dry ground. It gave a bigger head (the grain-bearing section of the tip of the stalk) and to his mind suggested the capability of producing a higher level of nourishment. Nothing startling had been achieved. But then nobody had the vast numbers of plants that he had visualized for crossing. The problem had never been tackled on a large scale.

There was another reason Borlaug needed to go forward. With the major work of the dwarf wheat behind him, what was ahead? Triticale was a very distant goal; an enormous amount of work would be needed. But if the prize could be won, the reward would be incredible.

He left Purdue at once and flew to New York to win money for his new idea. He told the officers at Rockefeller headquarters: "It is a very long shot indeed, but I have a feeling we should try."

Dr. Lewis Roberts, one of the veterans of the early days in Mexico, was involved in the Canadian work on triticale. After talking with him in New York, Borlaug flew on to Winnipeg, where Roberts had made an agreement with the

University to share work and data on triticale studies. The Canadians, in return for their cooperation, would get full access to facilities in the Mexican wheat stations as well as additional financial help from the Rockefeller Foundation.

Borlaug and Roberts shook hands in 1964, setting out on a path that would allow them to cross a genetic bridge that pundits said could not be crossed. They were on their way to the first man-made commercial cereal in history.

Thirteen

Dr. Swaminathan was enthusiastic about the semidwarf wheats, and in November 1964 he urged the introduction of a countrywide demonstration of their potential. His colleagues on an advisory committee of Indian scientists—a body that was responsible to the government for planning agricultural improvements—agreed to the suggestion. "We shall give our farmers windows that will look out onto a world of plenty," Dr. Swaminathan said, "a world of prosperity that awaits them through the use of science."

One reason the scientists were willing to adopt Swaminathan's proposal was that a beneficial monsoon in the summer of 1964 had readied the soil for fall planting, and the spring harvest was expected to yield 12 million tons—a 20 percent increase over the previous record. Oddly, the scientists were more receptive to suggestions of experimentation when the sense of urgency was lessened. Moreover, Swaminathan suggested only a modest beginning: demonstration strips of two and a half acres would be seeded on many hundreds of farms throughout India's 40 million irrigated acres. The farmers themselves would prepare the land and be responsible for cultivation and harvest after receiving the seed, fertilizer, and instructions. He had his critics.

They said that he was too ambitious, that farmers resisted such change, that this was a pipe dream. But he knew it was possible. He had run his own test on twelve acres of semidwarfs the year before, planting the Sonora 64 and Lerma Rojo 64 varieties that Borlaug had air-freighted to him. Using the two-season planting method common across India (growing wheat in the rabi, or winter season, and corn and millet in the kharif, or summer season), Swaminathan had gained a yield of ten tons, against a former average yield of one ton—a tenfold increase. He said later: "I believed the plots of semidwarfs would become the catalysts that would destroy indifference toward the use of science—at all levels in our society."

The plan was given the green light by the government. In essence it employed Borlaug's tactics of generating, through careful publicity, a hunger—even a greed—for the new seeds, new fertilizers, and new technology. In the wide, brown plains of India it went further than that; it caused a seed fever not known before or since in the subcontinent.

As the rabi season drew on in New Delhi in March 1965, Borlaug sensed an end to the good conditions in India. He felt a premonition of disaster, as he had in Mexico when the rust epidemics were looming. There was something in the air that made his scalp prickle. He summoned his colleagues in India and Pakistan to a meeting at his small hotel, near the Rockefeller office. They sat in cane chairs in the humid air as the punkahs whirred above them, waiting to hear what he had to say. A decade later Narváez recalled the scene with clarity: "I could see he would not spare himself—or us. He was full of urgency and zest to drive on quickly, to get more and more foods out of those lands. He gave this sense of urgency to all who would listen to him. He caught the ears of political and scientific leaders alike. Most of all, he caught the ears and hearts of those of us who worked in the fields, and we went about the work as though we were following him—like disciples."

Once again Borlaug was ahead of his time. He could feel the trouble emerging; he was apprehensive that government action might be lacking as a follow-up to the program. They would have to meet the hunger for seeds and for fertilizer that Swaminathan's "windows on plenty" would generate, and this meant importing fertilizer and seed. "It would be criminal and foolhardy, if not downright dangerous, to show these poor people this world of plenty and then deny them the ingredients," he declared. "We have to give them more than seed. We have to give them the whole Mexican crop production strategy, the whole technological package. It all has to be harnessed to sound government policy or the harvests are doomed before they are planted. We have to work hard in the political corridors as well as the wheat country. We've got to see that the economics are right, that the government announces its firm price for wheat *ahead of the season*. The little guys out there have to know they'll get a fair and stable price to pay the costs of the seed, fertilizer, weedkillers, and insecticides, and of their own sweat. They have got to know that they are the guys who will do the work. If they don't work, India will get nothing."

News followed fast from India that Swaminathan and his colleagues had been granted permission to purchase "some hundreds of tons of seed" from Mexico. Pakistan wanted some too. In this case the rivalry between India and Pakistan served their mutual agricultural interests.

It was a good start to the summer for the farmers of Sonora, the state from which the seed wheat for India and Pakistan was to be drawn. The two nations ordered a total of 600 tons of seed, worth 600,000 pesos, an important injection of capital into the rural districts. And when Borlaug returned to the Yaqui early in the summer, the grateful farmers welcomed him with a special ceremony. The local farmers turned out to thank him. There were speeches, and Don Rodolfo addressed him as "Mexico's friend." Borlaug was deeply moved, almost dazed. Roberto Maurer came forward and presented him with an envelope and a transparent plastic box on a plinth of black ebony. Set inside the box, carved with Aztec symbols and curved as though it were bending to some imagined breeze, was a head of Mexican dwarf wheat, beautifully sculptured in solid gold. Touched by the gesture, Borlaug stumbled through a little speech. He called on members of the CIANO group to stand and take credit with him, saying: "No man can ever do anything like this alone; it takes a great team." Then, holding his gold wheat spike in one hand and the unopened envelope in the other, he left the hall. As he walked past Dr. Villegas, she could see he was crying and could hear the small Mexican band at the back of the hall playing a traditional Mexican melody. She knew it well. The song was called "La Espiga"—"The Spike." She remembered the words of one verse, which told of heads of wheat in a field:

> The spike of wheat that ripened first
> Said to another spike nearby:
> "In our own land these grains will grow,
> In foreign soils our seed will die."

At the government seed commission on the other side of town bags were being filled with the grain of the best dwarf wheats—Sonora 64 and Lerma Rojo. Soon the 600 tons would be loaded on trucks and sent to Los Angeles, where they would be shipped halfway around the world to the foreign soils of India and Pakistan in time for the winter planting season.

Back at his hotel, long after midnight, Borlaug remembered the envelope Roberto had placed in his hands with the gold wheat spike. He opened it. Inside was a check for 50,000 pesos—about $5,000.

The telephone in the Maurer home awoke Teresa. She was astonished to hear Borlaug on the line. He sounded distressed and she at once woke Roberto. Borlaug was brusque, upset. He said: "I can't take this money. It's impossible for me to do that and still work here in this place." Roberto replied: "It is a gift from the hearts of many. How can you refuse it?" Borlaug said he could not look the farmers in the face if he took money from them. Roberto knew him too well by then. It was useless to argue. Borlaug said: "You'll just have to find some way to give them this money back." Then he hung up. Next morning the check was returned to Roberto and it was never discussed again.

In mid-August the Rockefeller office in New York asked Borlaug to take control of administration in Mexico City while Wellhausen took a month for home leave. For twenty years Borlaug had escaped spending time in offices; he had worked and lived in the open. The thought of spending weeks at a desk in an office in a heated, overcrowded city depressed him. He told Margaret it was like going to prison—only that would be quieter. He protested to Rockefeller officials in New York, asking them to find someone else. Harrar told him he had to do it. So he dressed in a shirt and tie and suit and did it.

As the new seed was being sent by truck from the Yaqui, Narváez flew from Pakistan to Los Angeles to supervise its loading on a freighter. The young Mexican scientist was worried on two counts. First, the tension on the India-Pakistan border could erupt into full-scale war, which probably would affect the seed plantings. Second, the seed had not yet arrived in Los Angeles. Sailing time was getting closer. Where were the trucks? Narváez telephoned

Mexico City to speak to an administrator; he got the temporary administrator, Norman Borlaug. Narváez was tough anyway: "Don't you people down there know we have to meet a shipping deadline? When will it get here?" Borlaug checked; the trucks had left two days ago. They should be pulling into Los Angeles any time now. He hung up, looking at the papers on his desk, a stack that got higher every day. Narváez was on the phone again the next day. He was getting more and more worried—still no seed. What about planting time? Borlaug checked. The trucks should be arriving any hour now. But the seed should really have been there long before. Where was it? Had thirty-five trucks been lost?

As it turned out, the trucks were in Los Angeles, but they could not get through to the docks. They were sitting on the side of a road that second week of August 1965 waiting for the Watts riots to end. There was no way for the trucks to get to the docks except by going through the riot-torn area. And as a Rockefeller Foundation administrator, Norman Borlaug found that he did not have the authority to end riots.

That same week, while India and Pakistan stood on the brink of war, and thirty-five trucks full of seed stood on the edge of Los Angeles, a telephone call brought more bad news to the administrator in Mexico City. The $95,000 draft from the Pakistani government in payment for the seed was irregular. It was marred by several misspellings, and the First National City Bank would not accept it.

Then Narváez was on the telephone again. He told Borlaug the convoy of trucks had arrived at dockside. He wanted the authority to go ahead with loading.

Borlaug's nerves were frayed. He told Margaret: "It got me in the stomach. I can deal with a rust epidemic, but this business of crisis and complication behind an office desk drives me frantic. It makes me unreasonable." Under strain, he had yelled into the telephone to Narváez: "Get that wheat aboard the ship—send it on its damn way!"

He told himself the check problem would be cleared up—somehow. Narváez, and his team in Pakistan, had a few days' grace. The freighter was hardly out of sight across the Pacific when war broke out between India and Pakistan, a full-scale shooting war in which thousands of soldiers were killed.

Two more crises followed. The Mexican seed commission, angry that there was no money to pay for 350 tons of wheat, made direct demands for payment on the Rockefeller Foundation in New York. Next, Borlaug was angrily involved in a complex deal with the President Shipping Line to divert the freighter, originally headed for Pakistan, to Singapore. He could not risk confiscation of the seed by Indian authorities at Bombay as material of war. At Singapore the wheat was to be loaded onto two other ships, one bound for Karachi, the other for Bombay. He had just completed this deal when his secretary broke in to tell him the Rockefeller office in New York was on the line. The official's voice was neither pleasant nor conciliatory. He told Borlaug that the decision to ship the seed when the check was irregular was "just about the worst administrative decision possible."

Borlaug exploded: "What the hell! I didn't want this blasted job—you ordered me to this desk. I don't want it now!" He hung up, fuming, then sent a cable in desperation to Malik Khuda Bakhsh in Pakistan asking for action on the check. Finally he said to his secretary: "No more telephone calls. I'm not here!"

He was at the desk trying to suppress his anger when the telephone began ringing again. His secretary said: "I'm sorry, Dr. Borlaug, but the Mexican seed commission and Rockefeller office are waiting on the line. There are also two press calls, and we expect another long-distance call from Pakistan."

He put on his jacket and walked into the outer office.

"Tell them all that Dr. Borlaug is sick. Try someone else, or leave it until Wellhausen comes back. Tell them all I'm ill in bed with influenza."

He hurried out of the building, feeling he had escaped from prison. The seed was on the way and there was nothing more he could do. Time would take care of it all. The world would try to influence the two nations to stop fighting. He went back to his home in Lomas, packed his rod and line and tackle, and went fishing.

Once again Borlaug's instincts proved to be correct. Time was the essential ingredient, but the next few months were filled with danger. Pakistan and India were at war, and Dr. Ignacio Narváez was working against the clock. The mishaps that had attended the consignment of seed to Pakistan had stolen weeks from the rabi wheat-planting season, and he had waited impatiently at the dock in Karachi for the old freighter to steam into port. With hard work, government support from the top, and special crews, the 350 tons of seed were loaded in railroad cars and sent to the north within forty-eight hours. The first planting occurred three days later. Meanwhile, a corrected bank draft was sent from Karachi. The predicament with Rockefeller headquarters and the Mexican seed commission was eased, but not forgotten.

Narváez started the program throughout the country at its northwest frontier, sweeping between Peshawar and Islamabad, working as far west as the foothills of the Sulaiman mountain range. Then he and his Pakistani colleague, Dr. S. A. Qureshi, went southward along the winding valleys of the great Indus waterway, planting in small plots on 2,500 farms. The aim was to fire an interest among the rural people. Seed was also sown by government men on fifty government stations, but many of the plantings were a disaster. The plots were not fed or watered as Borlaug and Narváez had stipulated.

Two weeks later Narváez turned back to the northwest frontier region, hoping to see the first sheen of the vigorous, short coleoptiles—the green spears of wheat that pierce the soil before the first leaves unfold. He saw very little green. Instead there were long, painful gaps between the green shoots.

What had gone wrong? The land had been well fertilized, properly cultivated, and watered, but the fields were almost bare. Had the seed of the wonderful dwarf wheats died in this foreign soil? His heart sinking, Narváez traveled on to more farms. Their hopes for wonderful growth, for shining examples of what was possible, for bringing about a rural transformation—all seemed dead. It was a terrible blow.

He rushed south to Peshawar and contacted Haldore Hanson by telephone; within an hour a cable was on its way to Borlaug in Mexico.

"We have been struck a cruel blow," the message said. Borlaug was stunned. The cable was a cry of distress, and he immediately flew north to Los Angeles. From there he was soon on his way to Karachi.

He was exhausted when he arrived, having worried himself sick on the flight, but still he could not rest until he knew the details. Hanson met him at the airport and wisely insisted he sleep for a few hours to prepare for the work ahead. As soon as Borlaug awoke, they started to the north to inspect the plantings.

As they traveled along the southern reaches of the Indus Valley on the rim of Baluchistan and northward into the Sind, the sorry tale unfolded before his eyes. The seed had germinated poorly everywhere. He drove alone to meet

Narváez at a hotel in the central town of Multan. When Narváez joined him, Borlaug saw his gray face and sad eyes and knew the worst.

"Nacho," he said. "You have seen the same horrors that I have seen!" He told Narváez of one meeting he had had with skeptical Pakistani experts. "I could see the smirks on their lips as they asked me whether these were the wonder seeds we had promised would work marvels. I wished the earth would open and swallow me up. What could I say?" He was overwhelmed by the predicament. He tried to stand back mentally, to survey the situation calmly and rationally. Excusing himself, he decided to walk alone in the early evening.

Pakistan is subtly different from India. Traffic flows on the left-hand side of the road, British fashion, but it is a mixture of traffic like few places on earth: buses, trucks, bullock wagons, a camel or two, and tall dark-eyed men in turbans bent over the shafts of carts piled high—always observing traffic rules amid the noise and dust. Borlaug saw the open doorway of an Anglican church. As he had done in St. Peter's in Rome, he went in, and taking a corner seat, his chin on his chest, he thought about the problem.

Narváez waited for him at the hotel, hungry and impatient. But when Borlaug came back, he said he could not eat. "Let's have some coffee and common sense, Nacho," he said. "Let's try to talk the guts out of this thing." So they sat on the balcony, drinking poor coffee and talking over the blow they had suffered. Later Borlaug rated that night in Multan as the most miserable in his life's work. The success of their campaign in Pakistan and their fight against hunger in India was hanging by a thread. "It seems more like a cobweb," he growled to Narváez once as they talked the hours away. Suddenly the reason for the disaster seemed clear.

"That damn seed had a jinx on it from the start!" Borlaug exclaimed. "It didn't come to life here, Nacho, because it was maltreated before it started. If my feeling is right, that seed was poorly handled back in Mexico. It was overfumigated and this burned the life out of it. We're to blame—no one but us. I got tied up with that darned office job in Mexico City, and we were in such an all-fired hurry to get that damn cargo out here, we failed to make post-fumigation tests! And that's where we went wrong, Nacho, sure as eggs. But how it helps, knowing that, I can't figure."

He calculated that four out of every five seeds had been killed. Fumigation was routine for wheat shipments, so as to kill destructive insects, but somehow the process had been carried to excess. The disaster seemed immense. How could he cope with it? The problem was not the value of the seed but the destruction of all that had been promised. Men who felt they had been misled, made to look like fools, would not forget easily or quickly. Borlaug was as near to defeat as he had ever been. Then he thought about the twenty years that had passed since he first camped in the old Calles research station in the Yaqui Valley. To give up the fight now was unthinkable. They would work on this thing until they dropped. First thing in the morning they would drive the hundred miles to Lahore and send a night letter to Mexico City. From there Wellhausen could advise Anderson in India that the only tactic left for recovery was to double the seeding rate on all remaining planting, double or treble fertilizer applications, double the amount of water, and hope the fertile seed would respond.

They met with Hanson in Lahore and talked with the Norwegian-born Dr. Oddvar Aresvik, senior economic adviser to the Pakistani government. He was Borlaug's ally. Aresvik had fought for the Mexican seed, and his persuasiveness had been partly responsible for the order of 350 tons.

Dr. Aresvik at once saw Malik Khuda Bakhsh; telegrams to all government agencies involved in the sowing went out within the hour. Borlaug and Narváez took to the road for several weeks. They went together, as they had in the early 1950s, when Narváez was the young and raw agrónomo.

Their quick action brought results. As they drove through the country, they saw the changes—subtle at first, then more visible. An ubelievable, astonishing response was occurring in the fields. One after another, the sparse Mexican-Norin derivatives thrust up new tillers; head after head would appear. More and more of the dwarfs grew into bushes. The gaps along the rows were filled as the genes responded to the extra energy given the soil.

"Feed those wheats! Feed them!" Narváez and Borlaug sounded the call up and down Pakistan. In places, they had as much as three hundred fifty pounds of fertilizer thrown onto each two-acre plot—three or four times the normal amount in America. Almost as they watched, the green tillers flooded out into the bare patches; the heads filled. The battle was won.

On his way home Borlaug went to see Anderson in India to find out how his battle had gone. Anderson said that Indian farmers had ignored advice to sow the semidwarfs only three inches below the surface; instead they had followed the custom they used with their own native wheats, sowing down to six inches or more. In addition to the weak germination, the three extra inches of soil had also resulted in poor stand in some places. But the corrective measures of heavier rates of seeding combined with heavier than programmed fertilization had saved the day.

Borlaug was back in Mexico for Christmas. He was elated at the revelation of genetic strength or recuperative power of his astonishing dwarfs. Talking excitedly to Margaret, he said: "We had our backs to the wall and there was nothing that would save us—except their genetic potential. There aren't any other wheats like them in the whole world."

Borlaug deliberately turned his back on the controversy that arose over the fumigation blunder, which became a political football, bouncing on into the coming seasons. Nor would he take part in recrimination; he needed to learn a lesson only once.

Early in 1966, Joe Rupert paid the Borlaugs a visit from Chile, and they went together again to Obregón and the Yaqui. Rupert was astonished at the town's wide, paved streets, the high-rise buildings, and the towering silos. "Gosh, Norm! This can't be the same place it was nearly twenty years ago!"

"This is part of the fruit, Joe. Repayment for the sacrifices and all the things we did together to get it started."

Borlaug had a mission in mind at that time for Rupert. He wanted his opinion on how rust disease might threaten India and Pakistan.

He told Rupert: "Those armchair critics who rarely get into a field are sounding doom all over the United States. They're claiming that we've opened the floodgates to disease in India and Pakistan. We've got enough troubles without their scares. I want them answered, Joe! I'd like you to go and take a good look."

Rupert made the journey to the subcontinent and came back with a favorable report. He could see no unexpected dangers. There were no disease threats of any unusual kind.

Within weeks of Joe Rupert's return Borlaug saw a new danger growing in India and Pakistan. The harvest of 1965 had been a record in both countries. India's wheat harvest was up a million tons, to around 12 million, and Pakistan cropped 4.5 million tons of wheat, half a million tons greater than any previous yield. Advances in basic science, improvements in irrigation, the boring of wells, and the efforts of extension officers among the farmers had all contributed. But essentially it was a rich gift of good weather that had lifted grain supplies and helped the two governments to conserve small buffer stocks against the leaner years. The good fortune produced political praise for the vaunted five-year plans, and in the euphoria of temporary plenty tens of millions forgot tomorrow, next year. Life was good.

There was also hope of an end to war with the coming Tashkent conference. In January 1966 Premier Aleksei Kosygin of Russia persuaded the soldierly Pakistani president, Ayub Khan, to meet the quiet and diminutive Indian prime minister, Lal Bahadur Shastri, for conciliatory talks. They met in Tashkent—a Russian city soon to be racked by disastrous earthquake—and signed an agreement to withdraw troops to ceasefire lines laid down by the United Nations.

Shortly after signing the agreement Shastri suddenly collapsed, and within hours the gentle leader who had followed the charismatic pandit Jawaharlal Nehru into power in India was dead. Out of the trauma of national loss and grief the massed millions became aware that their destinies were to rest in the hands of the first woman leader of a major nation. Mrs. Indira Gandhi—no relation to the mahatma, but beloved as the child of the first Indian prime minister—took all eyes in the opening weeks of 1966.

In those days of sunshine, before the monsoon was due, a new report from Norman Borlaug was dropped on the desk of Malik Khuda Bakhsh, Pakistan's central minister of agriculture. It said, in part: "A year ago I predicted that an increase of 50 percent in Pakistan's wheat yield could be attained within eight years." Now, Borlaug claimed, the performance of the various strains of dwarf wheats made it possible to forecast a doubling of the wheat harvest in five years—if the government guaranteed that farmers would get credit, price supports, and fertilizer supplies.

The minister decided to act. Alert to Ayub Khan's interest, he called a national conference of experts to discuss the pros and cons of the introduction of Mexican wheat and technology to Pakistan on a wider scale. He packed the meeting with government scientists and planners and the most vocal critics among academics and administrators. He invited members of the economic advisory groups—the US-AID people, the Harvard group, the staffs of the Ford and Rockefeller foundations.

At the time Borlaug was in Mexico. He knew at once how important the meeting would be, how Aresvik had fought with economic groups for simple recognition of the crucial role of prosperous agriculture in Pakistan's economy. He knew he had to be in Lahore when the conference opened, and he left immediately for New York, paying a brief visit to Rockefeller and Ford headquarters before going on to Kennedy Airport for his flight to Karachi. However, blizzard conditions had closed down operations at the airport and he spent the night dozing in a chair. It was eight the next evening before he was on a jet bound for London, where he waited several more hours for a connecting flight to Karachi. Anxiety built inside him as the aircraft droned across Europe and the Middle East, and his second night without sleep ended with his first glimpse of dawn over the subcontinent.

Karachi was gray with first light and clinging humidity as he walked into the terminal, his tired eyes seeking a familiar face. There was none. He ached for hot water, warm food, and a good bed. As he went through the gate, a small Pakistani wearing a peaked cap marked "FORD" stopped him. "Dr. Borlaug, sir? I am from Mr. Hanson. No time, sir.

Plane leaves for Lahore at seven. Soon Mr. Hanson, Dr. Narváez will come."

He was shown to a VIP holding room, and an hour later Narváez came, grinning and extending his hand; then Hanson arrived.

They were not going to be able to get to the meeting on time. The big debate was set for ten that morning. They would not touch down in Lahore until eleven. Hanson was not too troubled by that. He was more worried that Borlaug was under great strain from lack of sleep.

It was hot when they arrived at the agriculture ministry and entered the big conference room. The blades of the huge punkahs turned overhead and circulated the hot air through the crowded room. People thronged around an enormous polished table in the center of the setting; the volume of sound was certain indication the meeting had not started.

As Borlaug and his group entered, the buzz of voices died away. All eyes turned toward them. Hanson scanned the faces quickly. Some wore the look of inquisitors, grim and forbidding. Borlaug spoke to his companions before they walked to the table: "If they want my blood, I'll fight back," he said. "Don't you people get personally involved. You have to stay, and live, and work here. I can fly away afterward."

The chairman was a deputy minister whom Borlaug did not recognize. He rose, smiling and courteous. He looked around the room, waiting for a moment of quiet. "Gentlemen," he said, "the minister has called us together to discuss the import and basis for the report from Dr. Borlaug and Dr. Narváez which suggests that our wheat crop can be doubled in five years. We have been waiting to hear Dr. Borlaug speak on this matter. Dr. Borlaug is now here with us and, if you will give him silence, I will ask him to speak."

This was easy; even half asleep, Borlaug knew the picture. He had made all the pungent points before. He told them how Mexico had taken thirteen years to reach self-sufficiency; how Pakistan could double its grain production in five years and be self-sufficient in eight. The science was there; it only needed application. He sat down.

The chairman gave the floor to a tight-faced man who was standing on his feet waving a piece of paper as Borlaug finished. Hanson leaned across to Borlaug and said: "Chief axman. He's from Sind University, former director in a frontier province. Doesn't like the suggestion that you can do in a few years what he has been unable to do in twenty."

The official was a spokesman for the attacking force. He quoted wheat yields from 1914 through 1964 and then asked if this stranger from across the world were gifted with special insight not given to Pakistan's scientists for over half a century. Did Borlaug know the soils of Pakistan better than generations of its people? Was this offer really for Pakistan's benefit, or was it a triumph of salesmanship? He was sounding a knell of warning for his people. There was something alien in the importation of the strange wheats across the world; it was a danger to Pakistan, a costly danger. Look at the threat, he said. Short straws! Pakistan's animals would starve and the nation would be without the straw it needed for many purposes. And red grain? Who would eat red grain in Pakistan? It was foreign; it was alien. The plants would be vulnerable to diseases yet to be met; millions of lives would be thrown away in pursuing a mechanized technology that the peasants could never muster. Their lands would become dependent on fertilizer and chemicals that they could not afford.

Borlaug was exhausted. "My combustion point is low," he said to himself, "but I must not explode. I am too tired to snarl at this man." He saw Narváez growing red-faced with anger. He spoke to him in Spanish and told him to keep cool, not to speak. He saw Aresvik rise to his feet and wave his hands at the chairman. The deputy minister did not see him. He was trying to quiet several men who were calling out supporting remarks. Borlaug thought Aresvik's ears went blue with exasperation. The accuser was still shouting, waving his hands and his paper at them.

"You will have no compunction. You will kill our animals. Our buffalo and our cattle will starve with your wheats in this country. Soon we shall all die. Many of us believe these wheats will make our people sterile, and also our animals."

Borlaug rose to his feet, chin tucked in, bottom lip out. He looked around the room, brows knitted, forbidding. At once he had quiet.

He asked questions of the chief inquisitor. He knew the annual yields from 1914. But how much wheat did Pakistan import each year? The man did not know.

"You import now 1.5 million tons, including gift wheat from the United States. What color is that wheat grain? It is red. You say Pakistanis will not eat red grain. What then happens to the 1.5 million tons of imported red grain? Does America give you wheat to throw away? Is your president to choose between color and hunger? And what nonsense here is talked of straw? Is this a scientist who speaks? An agronomist who speaks? These wheats do not have a single long stem. They have many, many stems! What authority attacks these wheats who does not know that the Mexican dwarfs make more wheat straw, not less! You will have so much straw you will not know how to use it. Science must be used to help the people of Pakistan. If it is not used now to make this leap in production, you will deny history. Your agriculture is stagnating under this kind of weak science. You will stand still while your population overflows your resources, and then you will go begging to the powerful rich nations, holding out your bowls for scraps of food. And what science is it that opposes this forward leap? A science that says man's daily bread will make him and his wife impotent? How easy, then, to solve the problem of the world's exploding population. All we shall have to do is feed them bread! This is utter, hysterical nonsense, undeserving of an answer from a critical and scientific audience. I will say no more."

When he sat down, he was surprised by the warmth of the applause. He was pleased at his own control. He felt he had won a battle.

Hanson said to Narváez: "He was magnificent. Borlaug at his best." The deputy minister took Borlaug by the arm. The central minister of agriculture, he said, was waiting to give him tea.

Malik Khuda Bakhsh was smiling. "I hear those fellows gave you a very rough time, doctor. I am sorry, but we must let them have their say."

Borlaug allowed himself a little anger with the minister. "Such people are roadblocks in your country. They are a menace because they oppose change on a personal basis. We have a word for such people in Mexico—'scientific hatchetmen.' They will lie in ambush on the people who have to go out and do the work. If they were real scientists, they would be pressing for these seeds right now to save further waste of time."

Later Borlaug was heartened by a talk with Dr. Qureshi, the Pakistani wheat coordinator. He was optimistic about winning high-level approval for a commercial-scale trial of the Mexican seeds. Qureshi said: "We do import red grain, and our people eat it. We realize we must go for quantity just now, not quality. We have faults, Dr. Borlaug. We also like to dress in white shirts and shining shoes and send our juniors to do the work in the fields. We have no contact among our scientists. We do not have cooperation; we have cutthroat competition. Workers at Lyallpur jealously keep their results secret from other research scientists. The country suffers because scientists are selfish and do not exchange breeding materials. But maybe we can change. Nacho has given us the lead, and we will follow him." Borlaug had his first hint of Narváez's messianic role.

Narváez was to spend four years in Pakistan. He reorganized Pakistan's approach to wheat production and, in changing methods, brought the research centers into contact with one another. Wayne Swegle in his Ford Foundation report, said: "He moved programs and he moved men." When Narváez left, the Pakistani scientists showed sorrow for the loss of a friend and mentor.

The shadow of hunger darkened across India in the late summer of 1966. Borlaug's wheat coordinator, Glenn Anderson, reported: "We are under savage drought. The monsoon has failed, and it is certain that yields will fall so low that the government must use its fragile buffer stocks to meet the hunger."

When Borlaug reached New Delhi that fall, he went at once to see the conditions in Bihar and Uttar Pradesh. The misery was frightening, and India was badly in need of help. Swaminathan took him to see the minister of agriculture, C. V. Subrimaniam. The minister was very anxious but said that massive help was coming from overseas.

The major grain-producing nations were responding magnificently to the plight of the peoples of the subcontinent; convoys of ships were loaded generously with wheat from America, Canada, and Australia. The years 1966 and 1967 were to see the greatest food rescue operation in history. Enough grain was shipped to keep 60 million people fed—and it was free. A mountain of 15 million tons of grain flowed into India and Pakistan in those two years. So great was the response that port facilities and road and rail transport were inadequate to move all the grain to the stricken areas. But the famine was only alleviated.

Borlaug cautioned Subrimaniam: "We'll not cure the damn trouble by filling food bowls. If this is not to happen again, you will have to fill your fields with high-yield dwarf wheats and also use the new dwarf rice, which Chandler is breeding in the Philippines. They are the only hope you have to lift this wicked blight from your people."

Subrimaniam showed great courage in his decision. At a time when the land was in the grip of famine, when India's overseas currency was at low ebb, he decided to spend $2.5 million on 18,000 tons of semidwarf seed—wheat that would go into the ground, not into the mouths of the starving millions. Mexican seed would fill 700,000 of the more than 100 million acres of wheat land, and most of the grain produced from those acres would go back into the soil, for the rabi harvest of 1968. It was a brave decision, a bold step. It recognized that India had to do more than fill its food bowls by charity. Subrimaniam was a man of decision and action.

Borlaug was in Ciudad Obregón when the seed-buying mission reached Mexico City. Wellhausen telephoned the news and told him it was obvious that members of the mission knew about the Mexican fumigation blunder. They did not want to buy seed from the Mexican seed commission; they wanted to deal directly with the farmers, to see the grain in the fields. They were taking a plane to Hermosillo, the capital of Sonora, that very day. But it was late in the season, and Borlaug knew they would have trouble finding enough seed.

Borlaug drove the 150 miles to Hermosillo in blazing

heat and arrived in time to meet the Indian delegates. He helped them in their negotiations with the farmers, and within two weeks the Indians had their seed ordered. Borlaug was delighted to see men he had had to persuade ten years before to use fertilizer now operating confidently in the role of international wheat dealers.

There were no delays this time. The Indians did their own fumigating to kill unwanted insects. They chartered two freighters, and from the Sonora port of Guaymas the biggest seed purchase in history sailed across the Pacific. It was a record that would stand for less than a year—until an order surpassing that amount was placed by Pakistan.

Other orders for semidwarf seed were placed with Mexico that year. The USAID group in Afghanistan asked for 170 tons. From Turkey came another order for fifty tons. What did not appear on the official Mexican lists was the export of fifty tons of seed grown from one of the crosses Borlaug had made at CIANO, but which had "escaped" from the station before official release. Certain officials connected with this transaction later called it the "most illegal wheat in the world" and described its shipment to Pakistan as "smuggling."

This semidwarf strain, known at first only by its cross number, 8156, was among many produced from Borlaug's special plot in the Yaqui Valley, where he kept the "best of the bulls." The new cross displayed the qualities of very high grain yield, stiff, short straw, rust resistance, and a good response to nitrogen fertilizer, as well as a good grain that made a fluffy flour. Although it had not gone into multiplication for commercial use, Borlaug had presented samples of the 8156 seed to the first young wheat apostles, who had left Mexico for their homes in Asia and Africa in the spring of 1962.

Bajawa and Chaudry had taken a sample of the seeds—some with red grain, some with white—to Pakistan. These were later given different names: the white were called Mexipak; the red were named Indus 66. The seeds were grown in the plots at Lyallpur. By 1965, when the rabi planting season opened, they had multiplied to eight bushels, and sown into thirteen acres of land they had produced 1,100 bushels of grain. So well suited were they to Pakistan's needs that they were eagerly sought after in black market deals. Mexipak made the perfect chapati, the wheat pancake beloved on the subcontinent. Also, under intense cultivation, it could outyield the earlier semidwarf varieties, Pitic, Penjamo, and Lerma Rojo, by about 10 percent.

Mexipak caught the eye of the governor of West Pakistan, Malik Amir Mohammad Khan, who at once wanted to order fifty tons from Mexico. The Mexican authorities had not yet multiplied this seed, however, so the request was refused. Ignacio Narváez was in Pakistan at the time. He knew a certain farmer in the Yaqui Valley who was illegally growing the seed for multiplication. Narváez contacted Borlaug to ask his help in the purchase of fifty tons. People in Lahore were behind the deal, he said, and there was a check for $25,000 waiting for the farmer as soon as the seed was shipped.

Borlaug feared that Narváez was on dangerous ground. Unauthorized export of seed could cost him his position, both with CIMMYT and with the Mexican government. He told Narváez he could be summarily fired. Narváez replied: "Fire me if you want, but we need that seed."

Narváez was tenacious, and Borlaug knew it. Weeks later a representative for an American organization with responsibility in Pakistan turned up in Obregón and made a deal with the farmer. The representative had a few sample seeds with him when he went to see Borlaug. "I know what you are here for," Borlaug said, grinning. "But I don't want to be seen with you." The agent held out the seeds and asked him to identify them. He did, saying: "We call that seed by a different name now; it is Siete Cerros." It had been named after the seven hills that stand between Hermosillo and the sea.

The agent somehow got across the border into Arizona with the fifty tons of seed, and it finally reached Pakistan under the guise that it was feed for animals. Mexipak became very popular in Pakistan. But more than two years before, it had also come to the attention of the wheat workers in New Delhi. Anderson, Swaminathan, Athwal, and Kholi had proven its value and also had begun crossing it with other varieties. Concentrating on selections of the white seed, they developed a highly successful line that became known as Kalyan Sona—Golden Savior. It became a favorite with farmers, and soon surpassed the Pitic, Penjamo, Sonora 64, and Lerma Rojo varieties in popularity.

The wheat revolution was reaching a new stage of power, but there were still improvements and refinements to be made. The Sonora 64 and the Lerma Rojo, already introduced into India, were excellent wheats for disease resistance and response to fertilizer feeding. But they did not have the highly prized white and amber grains of some of the other strains. Swaminathan and an Indian colleague, Dr. George Varaghaese decided to carry out a sophisticated experiment to correct this deficiency.

They placed 500 of Borlaug's Sonora 64 two-gene seeds on a tray and subjected them to nuclear bombardment from a radioactive source that emitted gamma rays. They were seeking a mutation in the genetic reproductive material and scored a lucky hit on one of the seeds. It was not detected, of course, until after the seed had been sown and a new wheat plant had grown. What it revealed was that a rearrangement of the chromosome structure had changed the color of the seed from red to golden amber—producing a grain that ground to white flour, for the ideal chapati. They called the new wheat Sharbati Sonora, and although it had many desirable characteristics it subsequently could not compete with Kalyan Sona. The expanding team under Kholi and Swaminathan continued to reshape and improve the two-gene wheats in other crosses with native varieties. Sonora 64 was to reemerge as Sharbati Sonora, and another new variety, Sonalika, was identified and multiplied. The productive Lerma Rojo 64 would also have new cousins—Chotti Lerma and Safed Lerma, both white-grain wheats. Borlaug's wheat tree was growing fast—not only in India and Pakistan but in Africa, South America, and other countries.

Borlaug returned to Ciudad Obregón for the winter planting—and to sadness. Don Rodolfo Calles was dead. It was a setback for the entire Yaqui Valley and a deeply personal shock for Borlaug. He had last seen Don Rodolfo hale and hearty, as full of energy as in the earliest days of their friendship. Roberto Maurer told him the end had been sudden. He had developed an aneurism; an artery above the heart had swollen. The famous cardiac surgeon, Michael De Bakey had performed an operation in Houston in an attempt to save his life.

"Everything seemed to go well," said Roberto. "He was resting in bed, recovering, when—boom! He was gone."

Borlaug knew how much he would miss Don Rodolfo. His memory raked the past—heads of Austin wheat snapping in the breeze, Don Rodolfo's face and straight figure at the field days, his smooth handling of subscriptions for the establishment of CIANO. The fields of test wheats, the nurseries of new varieties grew because of Don Rodolfo.

Modern Obregón, with its 170,000 people, was busy and prosperous because of his work and faith.

Borlaug did not know then that the citizens of Ciudad Obregón would soon name a wide boulevard after Calles —and then another one after Norman E. Borlaug.

Fourteen

In 1967 India and Pakistan suffered their worst drought in forty years. Reservoirs dried up. Nothing grew. Millions of people went thirsty and hungry; hundreds of thousands died of starvation. Children who survived sustained irreparable damage from protein starvation.

Life became a horror in tens of thousands of villages. Ponds turned into silted murk, and the stench from wallowing water buffalo permeated the air. The agony bit more deeply into crowded India than Pakistan. In places where village crops remained in straggling clumps, clinging to life, hordes of rats broke desperately from cover to chew the stalks and devour the fallen grain. The misery was appalling. Early in 1967 the government declared the vast northeast section of India a disaster area. Famine became official.

The first such declaration in twenty years of independence was a humiliating confession of failure. Inability to control the situation was marked. Bags of wheat piled high at the docks and railroad depots for want of transport as the ships of the generous nations were unburdened of their mercy cargoes. The anguish of those months flowed into a political atmosphere of heightened emotion, into a sea of passionate resentment.

In this climate the ruling Congress Party, which had

guided India since independence, came under a hammering attack. Rioters roamed city streets. Embittered students stoned government buildings. Mobs were enraged and urged to destructive demonstrations by militant zealots. The years of war, the conflict with Pakistan, the danger of invasion by the Chinese at the northern border, internal fights, parliamentary squabbles, blunders in planning, the population squeeze on stagnating food production—all contributed to the mood of agitation. And against this backdrop the Indian government prepared for the general election of March 1967.

Side by side with these events—unnoticed amid the tragedy of widespread famine and the passionate denunciation of political campaigning—a marvel was occurring across 700,000 acres of India's wheat lands. The seed of the Mexican semidwarfs, which would fill 6 million acres in the next rabi season, was goldening on billions of dwarf stems, even as the people of India went to the polls. Too few of the players on the political stage knew that the agricultural revolution was being forged into a powerful weapon against famines of the future. The nature of the event was little understood, even by many people close to the rural transformation. The wheat marriages conducted by Dr. Swaminathan and his colleagues and the 18,000 tons of seed from Borlaug's semidwarfs were to be a bulwark against wider famine disaster in the coming years.

In mid-February Borlaug flew to New Delhi to inspect the results of the first massive planting of Mexican seed in India. Margaret was with him. They had met in Rome, where he had been attending meetings of the FAO. Margaret had traveled from Scotland, where she had spent a lonely holiday on a fruitless search for some of her family's relatives. They were to take a vacation, a long-delayed honeymoon, after he visited India and Pakistan. They planned to go to Bangkok, then on to Hong Kong, and Japan, and home to the United States.

It was to be a leisurely vacation, there was no need to hurry. The house at Lomas, in Mexico City, was empty. Billy was at the University of Oklahoma. Jeanie had been married the previous June to Richard Rhoda and was now living in Oklahoma, where her husband was stationed at Tinker Air Force Base, just outside Oklahoma City. She was happy, and teaching high school Spanish. Roberto and Teresa Maurer had traveled up for the wedding, and Roberto had remarked: "You see this lovely, blonde, blue-eyed bride and you never suspect that the Spanish will pour from her lips as it does."

There had been many relatives at the wedding, but Norman regretted that his parents were not present. His mother was now crippled with arthritis, and his father suffered partial blindness from glaucoma. After the wedding Norman and Margaret had driven to Iowa to visit his parents, and Norman was troubled by their condition. His mother was almost helpless, but mentally alert and bright. His father was deteriorating fast. They held a family discussion in Saude and made arrangements for the old couple to move into a modern rest home in Cresco where they would be comfortable and cared for. The farm was sold. His grandfather's old house and farm had been taken over by his cousin Sina, who had been his teacher in his final years at the rural school. She had redecorated the house and filled it with more modern furniture. Borlaug sat in the kitchen and remembered the fragrance of his grandmother's home-baked bread.

It had all gone quickly. Now, as they flew into New Delhi, there was an acrid smell of burning in the air, and people in the Indian capital seemed moody, sullen, resentful. Borlaug installed Margaret in a hotel and was immediately immersed in discussions with Anderson, Cummings,

Swaminathan, Kholi, and others. The tide of change again claimed him and swept him into the countryside. Hope for the belated honeymoon died rapidly. Margaret kissed him good-by and he left by car for Bihar and the Punjab. Then she packed her bag and caught a plane to Bangkok, following their holiday route—alone.

Borlaug and his companions traveled through green lands fed with water and lush with wheat, and through dry lands seared brown and arid with despair. Thousands of villages had formed little cooperatives, as the Mexicans had done, to strengthen community action. They had caught on quickly to the fact that the Himalayas had been storing water beneath their feet. American scientists, working through various philanthropic agencies, had developed techniques for tapping water from beneath the ground by tube wells—metal pipes driven into the land through which the dormant water was pumped to the surface to open new life. Underground was one of the great artesian basins of the world; layers of rock, aquifers in the geological strata, trapped the seeping liquid below the surface. In places these underground formations carried so much water that the surface quickly became waterlogged and salts gathered in the soil.

Borlaug saw the tube wells by the hundreds, with electric motors pumping away, the clear water running down channels to irrigate the dwarf wheats. While horror and starvation prevailed in Uttar Pradesh, fresh water flowed in the Punjab, Harrayana, western Uttar Pradesh, and Delhi state. Here the Green Revolution was beginning to flower. Hope bubbled as the dwarfs took hold, and thin faces filled with eagerness. Borlaug had seen the joy before. Yet the enthusiasm of the peasant farmers still astonished him. Whenever he got out of the car to look at a field of wheat, the incident with Carlos Rodríguez in the Bajío would be repeated. Adults and children would come running to gaze into his face, to touch his hand. A farmer would pull his sleeve, tug his arm. "Come, please sahib Borlaug. Come and see my field of wheat. Tell me what you think." He could not turn away, for another farmer would bar his path with blazing, delighted eyes. They spent many hours on many days among groups of village farmers, with Kholi translating. He said to Anderson one night: "Those stringy little guys are like all the others on the lands of the earth—wonderful people needing only a little help so they can get to work." On that same trip he remarked to Dr. Kholi: "There are two kinds of farmers in India now—those with water and those without water. Only those with water have hope."

The tour ran into weeks. Then, as they came back south toward New Delhi from the northern plains, Borlaug accepted an invitation to attend a promotional luncheon at one of the two tractor assembly factories in India. Diplomats and government administrators had been driven down to the factory for the event, which was intended to stimulate the use of tractors in Indian agriculture. There was a speech by the Polish ambassador, whose country had an interest in sponsoring the assembly plants. A few other people made speeches. Then Borlaug was asked if he would like to say a few words. Yes, he told the gathering, he had a few brief observations.

During the last month, he said, he had seen a wheat revolution in the fields of India. If it was given the strength and support of government action, it promised to replace famine. It must not be stifled or smothered by words that had no meaning. He raised his voice: "If I were a member of your parliament, I would leap from my seat every fifteen minutes and yell at the top of my voice, 'Fertilizers!' No matter what subject was being debated, I would always cry out, 'Fertilizers. Give the farmers more fertilizers!' There is no more vital message for India than this. Fertilizers will give India more food. And if there is no more food, there will be an exploding volcano beneath the feet of political leaders of this land. If you feed these soils of India, there will be no more crowds of tractors standing around this factory, or any other tractor factory that may be built. I give you this message. Water is so precious in India. Yet you waste the true effect of the water you have when you fail to feed the soil that feeds the plants. Water, yes, but fertilizer too. Only with fertilizer can you dare to hope." Borlaug believed that India had reached a fork in the road where the signpost pointed one way to the Green Revolution and the other way to recurring hunger and famine.

New Delhi was in ferment. On March 30, 1967, the Congess Party appeared to have suffered a sharp setback. The political stronghold was crumbling as a result of the swinging vote patterns. Borlaug arrived in New Delhi and went at once to find the minister of agriculture, C. V. Subrimaniam. He made an excited, dramatic entrance, flinging his arms wide, pointing through the windows.

"Do you know what is out there in your country?" he asked the minister. "Don't you realize what is building beneath your feet? You've got to take action. Prompt action! All those footling plans the gangs of economists have prepared. Tear them up. Throw them away. Start again and treble everything, everything—fertilizer, water schemes, seed multiplying, credit, factories. Multiply it all by three!"

Subrimaniam held up his hands to stem the flow of words, grimacing with mental anguish.

"Dr. Borlaug! Please—please. You don't have to persuade me." He waved a batch of newspaper clippings. "I know what you have said. But it is too late! It is useless to berate me. I cannot help you." He was in despair. "I have given my life for my people. Now I am blamed for the hunger caused by the drought. I have been thrown out. I have just had the news; it is ended. From tonight, I am out of office."

Borlaug saw Subrimaniam's hurt expression. This man who had found the courage to order the world's greatest purchase of wheat seed in the midst of a national food crisis had been rejected. Borlaug wanted to stay with him, to console him. But he was booked on the night flight to Karachi. Subrimaniam said: "Borlaug, this work must go on without me. You must go on doing all that you can. Before you go, there is a man you must talk to, so please wait." He picked up the telephone. "This man may not be the new minister of agriculture, but he is the most influential man left in government, next to the prime minister. He is at present minister of petroleum and chemicals—and he is also chairman of the commission whose plans you have been attacking."

Subrimaniam fixed the appointment for seven in the evening. When Borlaug said good-by and was almost out the door, Subrimaniam called him to a halt. "Borlaug! Speak plain. Be blunt and direct. Hit him as hard as you hit me, or you will never win."

When Dr. M. S. Swaminathan ushered Borlaug into the petroleum minister's office that evening, Borlaug already knew a good deal about the man. He had been told by his Rockefeller colleagues that he was a strong influence with Mrs. Gandhi and a man whose opinion was held in high regard by his colleagues. He was proud and personable, with strong leanings to the left, a man with principles and a social conscience that had led him to act at times to his own political detriment.

The minister was affable and smiling as he went through the polite formal greetings of Indian custom. Borlaug didn't wait to be invited; he began talking at once, rolling into a torrent of words. "Mr. Minister, I am also glad to meet you, but I have no time to dally or beat about the bush. I have to leave on the night flight and I must be emphatic and outspoken with you; I must say some things that you will not like. But I am a friend of India and a friend of all your farmers, and I would not be true to you, to them, or to myself if I did not say these things to you." He leaned across the desk, looking into the minister's astonished eyes. "Tear up all those plans they have fed you with, everything you have reckoned on—fertilizer, money for credit, pumps, tubings. Throw them away. Start again and multiply everything three or four times. Then you'll be getting close to what is needed!" He could see the affront in the minister's eyes; he could sense that Swaminathan was taken aback.

The minister raised his voice to quiet him. Borlaug spoke back: "I will not be silenced, Mr. Minister! Your planning is atrocious! It is based on the flat and stagnant past. You have a new situation, new wheats that outgrow all the others. You have a chance to feed your own people for the first time! I *must* tell you what you have to do. I could not sleep with my conscience if I did not say these things to you. And you must keep faith with those millions of little men out there on the land. They have looked through Dr. Swaminathan's 'windows.' They don't know any science. They are superstitious, and some have thought it witchcraft when they saw these wheats growing at research stations. But now, Mr. Minister, they have seen the wheats growing on their own land! What you have in those little men are the best extension officers in the world. You have got to give them their chance, to protect them against the grain market manipulation sharks. Give them fertilizers and credit! If you don't do those things, then I tell you that you and your government are sitting on a volcano. The anger of these people will one day erupt and sweep you all away. And if that happens, it will be your blindness that has been at fault."

The minister shouted: "Listen to me a moment!" But Borlaug was not in a listening mood. In all of India he was the only man who could tell him these things, and he had to strike hard. "Imagine your India free from famine, free from the burden of importing food just to keep your tens of millions alive! It is within grasp. Imagine your farms filled with people close to the earth, but prosperous—a new stratum of your society, people with hope and dignity and new purchasing power, buying goods that your industries will supply. Food comes first, Mr. Minister. Then build the factories. Give them access to tube wells. Let them have the materials so they will have water when your canal irrigation services fail them. Come out in the open, bold and confident, and tell the farmers what they will get for their grain. Give them the chance!"

He hammered at the fertilizer supply question again and again. Be bold, he argued. Give the farmers fertilizer and let the little dwarf wheats spring to full life to feed India. It meant revision of policy, expansion of plans and ideas. It meant challenge to a tottering government. In the end he spent nearly an hour with the minister, talking, and later listening.

As they left the office, Swaminathan whistled through his teeth. "For a time there," he said, "I thought he would throw you out on your neck."

Borlaug went to the airport, and Anderson was there to see him off. Borlaug said to him: "Keep after them, Glenn, I have a feeling when they come out of their bewilderment over the election they'll be looking for the very thing we've been arguing."

The new regulations he sought from the minister would be implemented in the days ahead. But by then he would be in Lyallpur, involved in another exercise of high-level political influence.

The wide, green lawns of Lyallpur University held the sparkle of a state durbar under the brilliant blue of the Pakistani sky. It was a ceremony given by the president, Field Marshal Mohammad Ayub Khan, to mark the end of the conference on the future of agriculture in his country, and it carried all the color and pageantry of the great levees in the days of the imperial British raj.

Early in the day Malik Khuda Bakhsh told Borlaug that the president would like to speak to him at the gathering. Borlaug said of course. Malik Khuda Bakhsh delicately suggested that there was a question of attire. Borlaug smiled. He and Narváez had set for themselves, their colleagues, and their followers a standard of dress in which they took great pride. Their suntans, shirtsleeves, and boots were a uniform. They had gone to visit ministers, governors, and department heads in these uniforms. "This isn't disrespect," Borlaug explained to the official. "If we can get across the idea that there's dignity in work and sweat, we might promote change down underneath."

Malik Khuda Bakhsh was conciliatory. Pleasantly, he said it was for Borlaug to make the decision. "But you see, you might consider it more proper to put on your coat and tie in this case. I know it is against your principles, Dr. Borlaug, but this is a rather formal occasion. It is not like being in the fields." Hadn't the man understood a word Borlaug had said? Nevertheless, Borlaug put on his coat and tie and went to the lawn party.

He stood with Hanson, Narváez, and Aresvik. Looking across the sweep of lawns and through the color of formal uniforms and the sound of modulated voices, he saw the hidden scenes of a vast parade of villages and hovels without water and power, with millions living on meager holdings, struggling to raise crops—people to whom plenty and peace were strangers, to whose eyes this setting would be a dream from the *Arabian Nights*. Borlaug saw the meeting as an opportunity. The president wanted to speak with him. Now he had a chance to touch a mind that wielded great power in the country. There he was, in the center of his retinue, seated on a shaded rostrum. Urbane, regal in bearing, aloof, his khaki tunic spiced with the hues of ribbons and awards, Ayub Khan was the political master of the 100-million-strong Islamic nation of Pakistan, named from the Urdu for "land of the pure." Forged from the religious division of the subcontinent when the British departed in 1947, Pakistan had been born in blood and massacre. Ayub Khan took power a decade after the death of the founder, Ali Jinnah, and in the ten years that followed he had changed the structure of Pakistan's government with military exactitude. He was a soldier who claimed to be at heart a man of the land and who greatly prized his family farm, where Narváez already had planted dwarf varieties.

Borlaug felt all eyes on him as he walked across the grass to the rostrum.

When Borlaug sat down, by some unseen signal the rostrum was cleared, and he was alone with the president. Ayub Khan said to him: "Tell me the facts. I am a soldier, but remember that I know farming and that my country has a deep weakness in a backward agriculture." Borlaug took the invitation to make his case. Pakistan could double its wheat output in five years, with new seed and a new technological package. Ayub Khan asked him what was needed. Borlaug used a military analogy: "You have a situation here where the artillery won't talk with the cavalry, and the cavalry will have nothing to do with the infantry. The battle will never be won with fragmented attack." He told Ayub that the scientist and farmer must work together. His government should strongly support the small farmer, giving him guaranteed grain prices, plentiful supplies of fertilizers, and access to credit.

They talked for an hour. Then Ayub said: "I have neglected my guests far too long. Please set out the details you have told me on paper—as simply as possible, for I am a plain soldier. Review all we talked about and have it sent to me in Karachi, just as soon as you can." Ayub rose and shook his hand. He looked serious. "Borlaug, we both have need of each other. You have my trust. If there is any difficulty, my door is always open to you."

That evening Borlaug's hotel room was converted into an office. The four men worked together—Borlaug, Narváez, Hanson, and Aresvik. The bed was used as a desk and they scribbled their notes on hotel stationery. Borlaug said: "We have the ear of the supreme power in Pakistan. We can set a shining example. Let's tell him the facts, hard and simple and brief."

He was concerned about his Mexican comrade, Narváez. He was still officially head of Mexico's wheat research, on loan to Pakistan through a Ford Foundation grant to CIMMYT. The job before them would hold Narváez here for at least another two years. How did he feel in taking up this mission?

"Nacho," he said. "You are close to this thing here, and I know you feel it's possible to grasp the achievement of making Pakistan self-sufficient—as we did in Mexico. I want us to achieve it before I fall over with a heart attack and hang up my working boots. How about you?"

Narváez made clear the depth of his own commitment, and the four men turned their energies to the memorandum to President Ayub Khan.

When the document was typed the next day, it was only two pages long—thanks to Hanson's journalistic economy. With a brief covering letter, Hanson delivered it to Ayub Khan's residence in Karachi the following day.

Borlaug and Narváez went out on the road again, visiting the research stations at Peshawar and Tendojam. They saw young men newly returned from the apostle school in the Yaqui. They were selecting out new wheats for further cultivation in some of the plots at Tendojam; Borlaug and

Narváez watched. One young argonomist looked at a small group of bent plants. These were good yielders, he said, early types, with rust resistance. But they were bent and that showed a tendency to lodge; therefore they must be discarded. Borlaug smiled and put his hand on the shoulder of the young Pakistani. "Stay a moment, please. In the Yaqui Valley we tried to teach you that if you listen carefully the plants will talk to you, tell you things. These plants are talking to me. This wheat tells me that two dogs came here to find a little privacy for their personal affairs. The wheat is bent over, yes. But let's call it social lodging. Give these wheats another chance. Grow another generation."

In the summer of 1967 Pakistan ordered 42,000 tons of semidwarf seed from Mexico. Ayub Khan had sent out his edict calling for special effort and unity; arrangements were being made to increase fertilizer supplies. A seed-buying mission was sent to Mexico.

As the Mexican farmers pored over ways and means of meeting Pakistan's enormous order, further requests for seed came in. Among them was a second order from Turkey, this time for 22,500 tons of semidwarf seed. These orders put nearly $8 million into the pockets of Mexican wheat growers. Borlaug took no active part in the negotiations for the record-breaking wheat seed sale, but he did witness the final signing at a new motel in Obregón. Afterward he telephoned Narváez in Lahore and suggested the text of a new report they would send to Pakistani officials. "Tell them we now revise the forecast of six months ago that Pakistan can double wheat production within five years—that we go further. Say to them, in writing, that it is technically feasible now for Pakistan to achieve self-sufficiency in wheat *and to go into surplus* in the coming harvest of 1968, and any year thereafter—providing the government keeps faith and meets the demands for fertilizer and the extension workers fulfill their assignments."

Borlaug was satisfied that the peoples of the underfed world had a good chance of escaping a repetition of the famine they had known in 1966 and 1967. Filling bellies, however, was not enough: bellies had to be filled with nutritious substances. The rising price of meat was placing animal proteins completely out of reach. Science had to make new advances in nutrition.

Borlaug was encouraged by the CIMMYT work on Opaque 2 corn. And across the Yaqui fields was another hope: the thousands of triticale plants his team had been crossing and experimenting with for the past few seasons. It was time, he decided, to give this work a vigorous push.

He caught a plane out of Mexico to New York and from there flew to Ottawa, Canada. He had a man in mind to take charge of the triticale development program.

Dr. Frank Zillinsky was surprised when Norman Borlaug called to see him. The big, broad Canadian scientist had a reputation for his work on oat plants, and he and Borlaug had met many times before. Borlaug thought Zillinsky had the strength of character and the physical stamina needed to expand the triticale program. Zillinsky knew about the scattered experiments done to date on triticale. He knew Muntzing of Sweden, who had done some of the original work; he knew that further attempts to produce something worthwhile had been made in Spain and by Dr. Len Shibeski at Winnipeg. He was cautious, hesitant. He understood a plant breeder could get sidetracked into a single pursuit and spend his life's work on a quarry that could never be captured.

"Why me?" he asked Borlaug. "What do you think I can do that other people have not been able to do?"

Borlaug said triticale should no longer be a curiosity. It

should be a major objective. The Purdue discovery of high-lysine corn from the mutant Opaque 2 gene had opened a door to higher nourishment with grain-based protein. He believed—in fact, he had some evidence already—that rye-wheat crosses would deliver a higher balance of amino acids, producing higher protein content, greater yields, better nutrient values.

Zillinsky pointed out that no matter which way the rye-wheat crosses had been made, they had always come out highly infertile. They kept running around in the same circles. Why, then, would it be any different in Mexico?

Borlaug answered him. In the Mexican fields were the plants, the space, and the army of technicians to work on triticale on a scale never before attempted.

"Big numbers, Frank," Borlaug said. "We shall turn that wheel of fortune a million times. And one day we'll win a gold nugget. Due respect to Muntzing and all the others in triticale work. But no one has ever played it big. That's what I want you to do. Will you come?"

Zillinsky had three weeks' leave due later that summer. He said he would spend his holiday among Borlaug's triticale, in Toluca Valley near Mexico City. Borlaug was certain that once Zillinsky was face to face with the prospect of breeding the first man-made cereal, he would be captivated by the task.

The semidwarf wheats from Mexico fattened their ripening, plump kernels by the billions in the spring sunshine of the Indian subcontinent in March 1968. They were building to a new record harvest in both India and Pakistan. Neither country could fill more than a fifth of its wheat-growing areas with semidwarf seed, but the dwarfs produced 42 percent of the total yield. India harvested a crop of 16 million metric tons, while the Pakistan harvest reached 7 million.

It was in the average yield per acre that the semidwarf performance was most marked. The advance was best seen in comparison with the best tall Indian wheat variety, known as C306. The C306 crops were fed with 50 pounds of nitrogen fertilizer per acre. Feeding above that ratio caused the tall wheats to lodge. But the 9 million acres of stiff-strawed semidwarfs grew short and level as if they had been trimmed to even length—and they could be fed up to 275 pounds of fertilizer per acre. The tall Indian variety gave an average yield just below fourteen bushels an acre, while the average yield of the semidwarf varieties was above thirty bushels an acre.

Borlaug had seen the crop ripening that March as he traveled from New Delhi to view the millions of acres across the Ganges plains. Then he crossed over into Pakistan, where he toured the mighty Indus Valley from the northwest frontier to the lush fields of the Sind.

Narváez was a happy companion. Together they sensed not a successful conclusion to their work but the beginning of a new era. Borlaug stopped in Lahore to write a personal report to Ayub Khan. He gave it the title "A Target Reached": "I am confident Pakistan will reach—or at least nearly reach—self-sufficiency in wheat production during the current season."

The goal had been attained two years ahead of schedule. The farmers had accomplished within three years what it had taken thirteen years to achieve in Mexico. "Mr. President—across the wheat fields of your land the new wheats have caught the fancy of your farmers."

They had also caught the fancy of many seed merchants. As in India, semidwarf seeds were hotly sought on a thriving black market. Prospects of quick fortunes turned some farmers into seed producers—and seed producers and farmers into thieves. Narváez reported that Mexipak seed had changed hands, at rates equal to $750 a ton. For more advanced varieties the price skyrocketed even higher. In both countries stealing from research stations and experimental plots became commonplace.

Raids and holdups became so routine that armed soldiers were put on sentry duty at many experimental plots. At one university research station, a gang of men overpowered the guards. While the sentries were held at knifepoint, the men raked through the dwarf lines for the most prized wheats, those with three genes for short straws. These promised higher and higher yields, but they were not yet ready for farm use.

The raiders escaped north to grow their stolen seed in the foothills of the Himalayas. They gained nothing. The strains they had looted were infertile.

In Pakistan Narváez heard of a stolen head of three-gene dwarf wheat selling for $14 a gram—a trading price of more than $6,000 a pound. In that same season Borlaug noted in his reports to the two governments that a similar fervor was greeting the introduction of the new rice strain IR8, which Chandler and his associates had developed at the institute in the Philippines.

Borlaug could not entirely deplore the fervor. It had a beneficial aspect. "It puts a bomb under complacent scientists," he noted. "They can't stand and preen themselves like peacocks when the sky is falling on them."

He found the advances in India exhilarating—though India was not yet approaching the self-sufficiency of Pakistan. The prime minister, Mrs. Gandhi, had asked to meet him. He was to go to her office in New Delhi for tea.

The Indian leader greeted him with charm and grace. She quickly showed a penetrating mind. She asked for his version of what was happening with wheat. "I hear so many things in this office that I have to take with reservation," she said. "I hear claims made in parliament and have to try to judge who speaks truly amid the wrangle and the ax grinding. What actually is happening with wheat production in this land? What do you anticipate will happen?"

"Madam," Borlaug said, "you have to face hard facts. You have a harvest coming in of at least 17 million tons of wheat. But that is only the first faltering step."

He spoke of the need for strong government action, and he forecast that India's grain output could double in a few more years. "I plead that you recognize the dynamic strength in your new force of scientists. Give them support to go out and to lead your farmers to produce more food. We have been accused in this land of being the enemies of the bullock and the buffalo and the cow. We have been smeared by people with interests vested in the manipulation of hunger. The government must take the lead and show the faith. If you do, I predict that by 1970 India will rank as one of the three great grain-producing countries in the world, behind only the United States and Russia. You have the capacity, the potential; you need only the will.

That was the challenge. But India could be maimed in these efforts by ineffectual population control, by too little fertilizer supply, by lack of proper water storage.

Human welfare in India always came back to water; drought was always a prospect. Fortunately an astonishing number of tube wells had been sunk across the subcontinent. "It is now known," Borlaug told the prime minister, "that each year we can draw off enough water to cover 4 million acres to one foot deep without lowering the level of the underground storage of water. That is the rate of recharge, the rate at which the water flows from the great mountains to the underground strata of your country."

And every gallon of water would be needed, he felt, when the summer brought its searing heat.

When Borlaug's plane landed in Obregón, Frank Zillinsky and George Varaghaese were waiting at the airstrip, grinning broadly under their straw hats.

"You're coming with us, Norman," Zillinsky boomed. "Come right this minute and see what we've caught!"

Without waiting for his bag, Borlaug hurried into Zillinsky's spattered Chevrolet, and soon they were rolling over the long, flat road and the ten miles to CIANO. He looked at his massive colleague, brows raised with the unspoken question. Zillinsky grinned. "Wait and see."

Zillinsky had been a dynamo since he joined the CIMMYT work on triticale the previous year. But Borlaug had many reports from workers in the field—none of which suggested a breakthrough. It was hard to guess a reason for this excitement.

Still dressed in a suit and lightweight shoes, Borlaug followed the Canadian scientist through the rows of plants, plot after plot, block after block; once he had to leap an irrigation ditch filled with muddy water. Zillinsky stopped in front of him, then stood aside. He stabbed a finger at the array of plants. "Look at that, Norman. Tell me what you see!"

There were many rows of the tall, heavy-headed triticale hybrids—half wheat, half rye. Big and clumsy, they numbered many thousands, of differing height, stance, shape. Among them, like straight lines drawn on a rugged landscape, were seven rows of plants, each one alike—trim, neat, uniform. Zillinsky was sure they were completely fertile. Each plant had all of its florets; the grains were all developing normally. They had crossed the bridge the experts said could not be crossed. The Yaqui fields had given the world its first completely fertile rye-wheatcross.

Borlaug went through the seven rows of plants with probing fingers while Zillinsky and Varaghaese looked on, confident of their confirmation. The first little family of fertile triticale. Was Zillinsky right? Had they actually produced a new cereal grain—the first in 10,000 years of agricultural history? The accumulated experience of decades told him that they had. Patterns of cells he knew so well were at work with others in these unique plants, developing along lines dictated by a marriage of genes and chromosomes that had never come together before, matching and merging—ready for reproduction. What had happened?

His eyes, his fingers told him the answer. Zillinsky stood waiting. Borlaug straightened, wonderingly. "Frank," he said. "One of those darned dwarf wheats jumped the gap!" He counted the number of fertile plants. The arithmetic fitted his picture. He recorded it more formally, later: "This fortunate combination came about almost certainly —and probably in the dark of night at CIANO—in March 1967. It was a natural, promiscuous—unplanned—outcross between a dwarf wheat and sterile, lank, poorly adapted triticale, one of those that we had derived from the Canadian material."

They sat for hours going over the records to trace back the event, to recall the dwarf varieties that were growing around the triticale plots. Borlaug said: "There was enormous luck there, Frank. It was one seed that grew from that night encounter. We were lucky to put it back into the ground. Your hand was guided. We must be thankful."

Zillinsky and Borlaug were full of urgent drive. They started a new breeding program. The new plants were in-betweens, their future behavior unknown. The requirements, though, were very clear-cut. Zillinsky wanted good things from both sides encapsulated in the new genetic marriage. But the Mexican wheats would have to provide the genes for dwarf straw; that was vital if the new cereal was to be a high-yield crop. Genes for dwarfing, genes for profuse tillering—all the other things—had to be mixed together in the fertile triticale choromsomes. And that amounted to a daunting, challenging plant development program.

In breeding the first generations of fertile triticale, Zillinsky struck difficulty. The crossing of desired traits into the triticale was more taxing than in wheat breeding because the genetic architecture of the new plants was still poorly understood.

Later that summer, when Borlaug was lecturing in Australia, Zillinsky worked through the blocks of rye plants in the high mountain plots at Toluca, on the other side of the road from where Borlaug, Rupert, and Richard Spurlock had cut the Supremo in 1947. Zillinsky recalled: "I went over those plants looking for good ryes. I just looked through them. I didn't read them by pedigree the way that Norman does. He was away, and so, rather than leave them to the trainees, I went through them. I saw a gap between the crowded rows and I pushed the plants aside to see what had happened. There it was. Quite by chance—small and sturdy. My first-generation dwarf rye. I called it Snoopy, for trying to hide from me."

He collected the seeds from his dwarf rye and look them to the Yaqui Valley. "This fellow made all the difference," he said. "I got really good dwarfism into the next generation of triticale and it made a big leap possible. It saved me years of frustrating work, I'm sure." There was little doubt in Zillinsky's mind that they would surmount future stumbling blocks. "Borlaug had seen the crack of light under the door. Now I could see it streaming out, and I was convinced we were going to win the world a new type of grain plant, a totally new grain—not rye, not wheat, but a cereal with new food potential."

The swirl of new work and activity surrounding the triticale project became an integral part of the CIMMYT fight against hunger. It was a battle that had been dramatically broadened as the precarious world balance between food and people became more fully realized. The equation was a simple one: world food production had to be doubled within twenty-five years to keep mankind at its present low level of nourishment. The Rockefeller and Ford foundations had been joined in the battle by the Kellogg Foundation and the Colombian government. New organizations had been established: CIAT, the International Center of Tropical Agriculture, based in South America; and IITA, the International Institute of Tropical Agriculture, set up in Nigeria under the sponsorship of the Ford and Rockefeller foundations and various government agencies. All this was linked with CIMMYT.

Fifteen

Norman Borlaug was in Australia in August 1968 locked in a motel room, writing a lecture that he had been invited to give during the third world wheat genetics symposium. He had enough facts now to talk for the first time—in public—of the Green Revolution in Pakistan and India and of the breakthrough in food production he saw coming in Turkey, Afghanistan, Tunisia, and Morocco. He had delayed the writing of his lecture until he had the latest reports from Anderson and Narváez. They had arrived only a few days before the symposium began.

He asked his colleague, Australian-born Dr. Keith Finlay, now director of CIMMYT's special projects division, to lock him in the motel room until the lecture was written and to open the door only to pass in his meals. He slipped out the written pages under the door, and he was in seclusion for three days.

He called his lecture "Wheat Breeding and Its Impact on World Food Supplies." On the night of the speech he faced the leading men in his discipline. He knew there were critics, many of whom would disparage and disapprove of what he had done. It was not uncommon among scientific audiences. That did not alter what he had seen, or what he had to say:

"The wheat production revolutions in India and Pakistan were not accidents of nature. They did not just happen. Rather, they came from a five-year struggle that culminated during the last three years in the development of accelerated production programs aimed at revolutionizing practices in both countries. . . . Today I am optimistic about the outlook of food production in the emerging countries in the next two or three decades. We have demonstrated the feasibility of shortcuts to increased production—if such attempts are properly organized and executed by skillful and courageous scientists."

He answered suggestions, made by some American critics, that weather had been responsible for the increase in grain production in the two recent years. "Weather has nothing at all to do with this success. In fact, it hampered our work, as the coming good years will show. What has been gained comes from sound, patient research done over long years—research that was not self-evident, since much of it was done half a world away in Mexico." He demonstrated how the increase in food was linked with fertilizer production. Fertilizer was a barometer in India, where the annual consumption had risen from 60,000 tons in the 1950s to above 2 million tons; it would need to go much higher to defeat hunger.

Finally, he criticized the scientific community. Hundreds of millions of dollars had been poured out by the advanced nations without result, because far too many scientists stood aloof while hundreds of millions of human beings struggled for survival. Scientists should function as catalysts for new ideas and methods instead of chasing academic butterflies. "We may afford such luxury in the rich countries, but for the underfed countries it is deplorable, ruinous, disastrous. I have observed the effects all too often."

After the lecture many of the scientists deprecated Borlaug's remarks. One international figure in the wheat world, who himself had once held a production record, was openly derisive. "Risky and impetuous," he called Borlaug's suggestions. "The man has invited epidemics," said another. Yet another said that Borlaug "displayed a typical national characteristic: boastfulness."

In the minority was Dr. John Falk of the Australian government's plant industry research organization. The world had been told of a new age, he said, and Borlaug had shown courage in words "written without guile or pretense and only in an attempt to present fact and outstanding achievement."

The sting of Borlaug's words lasted long after the symposium ended and the scientists had gone home to their different countries. Many made statements to explain away the mushrooming grain yields in the struggling lands, ignoring the long years of work in building defenses into the new wheats. In the United States there were many bitter attacks on Borlaug. The agricultural writer and consultant, Dr. William Paddock, had published a book with the startling title: *Famine 1975! America's Decision, Who Will Survive?* the theme of which was that there was no hope of agricultural research feeding the world's expanding population. Furthermore, the U.S. will have to decide which countries it wants to feed—on the basis of their political relations with America.

Soon after Borlaug's speech, Paddock was quoted in the press as saying that the wheat explosions were "illusory and caused far more by good weather than by good husbandry," that "the miracle grains of the so-called Green Revolution were of lower nutritional quality than ordinary varieties," and that the disease resistance the dwarfs possessed would turn out to be a "sometime thing." Paddock also took pleasure in quoting Dr. Paul Ehrlich, advocate of zero population growth and author of *The Population Bomb,* a book that predicted that hundreds of millions would die of starvation in the mid 1970s "in spite of any crash programs." Ehrlich believed the battle was already lost when he wrote the words. Paddock quoted him as saying, "Those clowns who are talking about feeding a big population in the year 2000 A.D. with make-believe Green Revolutions should learn some elementary biology, meteorology, agricultural economics, and anthropology."

There were also claims that the Green Revolution would lead to greedy land grabs, make the rich richer and the poor poorer. Borlaug labeled all the attacks preposterous. He told Paddock, by letter, that his writing implied that millions should be allowed to die of starvation. "We have bought you, and the other vocal gentlemen, time to deal with the burning questions of population pressures. You are barking up a wrong tree. Why don't you and the other talented writers turn your efforts to health problems, population, educational and housing needs, transportation shortages, and the real—not imagined—environmental deterioration, rather than attack genuine attempts to defeat hunger?"

The unwillingness of scientists to acknowledge the success in India and Pakistan made skeptics in the universities of the West. Confronted by accusing academics, Borlaug said: "If you cannot, or did not, go to these lands to see the misery and the thin children, then I say to you that you should learn from those who have. When you give more to eat to millions who have always been hungry, when you give more money to those who have always been poor, you create new energies, new factors. You stir up new thinking. A powerful new human force is set into motion." He conceded that this new force created new demands. Governments would have to improve their services. There were demands for better roads and homes, for power supplies, and fresh water, for medical services, education, and transportation. He said: "Is it suggested we should deny people food because they will want a better life? Can we solve in a wink of the eye all the social ills that have beset man since Adam, because we are growing more food?"

In 1969 India, originally the "slow runner," launched the largest national wheat research program in the world. It made the programs of the United States, Russia, Australia, and Canada seem modest by comparison. As Mrs. Indira Gandhi told her people in that year: "We are on the threshold of self-sufficiency in grain food production. The technical advances in agriculture should enable us to meet our rising demands. This year our total grain production will cross the 100-million-ton mark for the first time."

The revolution was building even faster than Borlaug had thought possible. India was producing more than 60 percent of its current fertilizer needs and was aiming at 3 million tons for use in 1974, with a total annual grain target by then of 129 million tons. The projected use of nutrients for the depleted soils was fifty times what it had been when Borlaug first came to India.

The Indian government had backed new fertilizer and tractor factories. The demand was without precedent. Assembly plants could not turn out farm machinery fast enough. In 1969, 35,000 small tractors were imported, and there was a waiting list of two years for new tractors. Farmers formed village cooperatives to sink wells and provide electric motors to pump the water to the surface, for both irrigation and domestic use. In the previous year a total of 200,000 wells had been sunk. For Borlaug, the thousands of miles of steel tubing that bit into the ancient soils were symbolic of a shattered myth—that the small farmer was an unchanging stoic. An official Indian publication dealing with this subject declared: "The Green Revolution has rightly demolished some cherished notions about the Indian farmer. He was generally regarded as apathetic, tradition-bound, averse to change. . . . The events of these past two years have shown an innate resilience and mental flexibility in our farmer. His outer crust of indifference, his resignation to his miserable lot were born of centuries of neglect, during which his travails were of no concern whatever to rulers or the intelligentsia. Such research as was done for the previous fifty years was divorced from the realities of agriculture and its burning problems."

However, there were still bleak spots of hopelessness in India. They would remain until India could take more positive action to improve the water supply or to step up the breeding of drought-resistant crops. But what was happening in the rest of India gave promise of greater hope and a new economic base from which these goals could be accomplished.

Early in March Borlaug told a meeting of economists that he thought the water revolution created by the tube wells was of greater human value than they realized. Not only did the wells provide nourishment for the crops; they also provided the villagers with a supply of clean, fresh drinking water. For all of history Indian villagers had used the water from local ponds. As a health step, the wells were of incalculable value.

"If water brings a new gleam to the villager's life, think what other things are happening in his world. Just consider the little guy in Bihar or western Bengal—one of 100 million without education. He's been working this same small patch of land for all the days of his life and it has given him no more than eight bushels of grain an acre, even in the best years. Suddenly, with other millions, he can get back seventy or eighty bushels. His return is up tenfold! His crop, like most of those on the 35 million acres of wheat fields in the land, has always been cut with a hand sickle, and the bullock has been used to go round and round to stamp on the cut crop and thresh it for him. Now he has ten times as much! How can he cut it by hand? How can that animal go round and round ten times as long? It will be on its knees while the winnowing goes on! This is the brief, dry period at the end of the rabi, and this guy can smell the rain in the air. The big black clouds of the monsoon are building up in the sky over his head, and he is a doublecropping man and wants to get his summer crop of millet, sorghum, corn, or rice into the soil. So—what to do? He now has money. He creates demand for little tractors that will help with the plowing. He also wants help with the threshing, and within weeks little blacksmith shops over India are breathing sparks from the fires fed with bellows. Soon there will be more thousands and then tens of thousands of small threshers rattling away across the land. Other farmers will put their money into cooperative schemes for machinery and hire their equipment out to the man who has a tractor but no thresher. With income there is new hope. With radio there is new contact and awareness and education on how to manage crops, how to preserve and guard their health. Political leaders will become names known to them. Suddenly knowledge has a new currency in India."

New wheat varieties from the Indian breeding program were further brightening the picture. The amber-grain wheats, Sonalika and Sharbati Sonora, were being rapidly multiplied on a commercial scale. And the Indian work was showing hope of winning even higher nutrition from grain. The scientific corps was also concentrating on a new high-protein corn developed from the Purdue Opaque 2 gene discovery.

In mid-March Borlaug went to Pakistan, where President Ayub Khan was to present him with an award—the Sitara-i-Imtiaz; one of Pakistan's highest civil honors. The countryside was full of activity. Enthusiasm was at a peak in the wheat lands. Among the people working in Pakistan was Borlaug's handsome Yaqui friend, Roberto Maurer. Financed by a Ford Foundation grant, he was spending a season among Pakistan farmers teaching them new wheat techniques of planting, irrigating, and harvesting. His wife, Teresa was with him; but she was working in Lahore, helping at a rundown hospital.

Despite the improved mechanization in Pakistan, Roberto Maurer saw more antiquated agricultural techniques than any he'd ever known. People still went into the ripening wheat stands with hand sickles, grabbing handfuls of stalks and slicing through them with the blades. It took six or eight men twelve hours to cut an acre. Most farmers, working without tractors, would harness a bullock and set it walking on a circular track, dragging a heavy frame and stomping out the grain from the stalks. Then the wheat would be thrown into the breeze by scoopfuls to separate the grain from the lighter chaff. The process was ancient and slow. The gleaners still followed the harvesters—as they had done before Christ. The hired hands were still paid in kind. In most places a bundle of about twenty-five pounds of wheat was a day's pay. Yet, among all this, change was evident.

In the cities, in Karachi and Lahore, there was change of another kind. Borlaug saw signs of weakening government control. Marching students, militant leftist groups, grew more daring and more defiant in their campaigns against the established order. Borlaug, Aresvik, and Narváez drove through the streets of Lahore to the presidential palace, where in a formal ceremony President Ayub Khan was to present his award.

Borlaug was shocked by the appearance of the president, who had suffered a recent heart attack. Formerly upright and soldierly, Ayub Khan was a sick man, with deep shadows under his eyes and faltering speech. When he came to the presentation of the insignia, which he held in his right hand, Borlaug noticed that he used his left hand to lift his right arm across the desk. Talk was out of the question.

After the presentation, Aresvik and Borlaug lunched together at the Ford Foundation staff club. Aresvik was still an adviser to Ayub Khan, but contact had been very slight since the president's heart attack.

While they talked over lunch, they were approached by a high-ranking economist who advised the Pakistani government. The man's expression was worried. He came immediately to the point. Pakistan was in deep economic trouble. Cutoffs of aid and support, resulting from the conflict with India, had created severe problems. Drastic pruning of expenditures was imperative if there was not to be a financial crash. What would be their reaction—and their Rockefeller colleagues' reaction—to a proposal to reduce the guaranteed price support for wheat by 25 percent in the present harvesting season so as to help weather the fiscal storm?

Borlaug was so angry he had difficulty swallowing his food. Ten million peasant farmers who would soon be laboring to get the harvest in were to be cheated. They had kept their faith; they had made good their promise to Ayub to produce a record wheat harvest of over 8 million tons. Now the government advisers were about to break Ayub's promise. Borlaug exploded: "Damn! Use your common sense, man! You cut the price of that wheat and you destroy the government once and for all. You'll blow the government out of its seat. I am not an economist. I don't know how you can find the money to pay these workers for their wheat, but pay you must. You break faith with the 80 percent of Pakistan's population who are rural workers and you'll light a fire that will make the present unrest by the minority groups in the cities look feeble."

They talked and argued, but it did no good. The man

went away still poring over the figures on his piece of paper. Borlaug and Aresvik went back to the hotel to pack their bags to fly to a Ford Foundation conference in Beirut. Borlaug was increasingly worried.

On the first day of the Ford symposium, at noon, Borlaug learned that President Ayub Khan had resigned. He would be succeeded by the commander in chief of the armed forces, Field Marshal Yahya Khan. His resignation opened troubled years for Pakistan, during which East Pakistan would secede and create the new state of Bangladesh. It posed a dismal future for the Green Revolution.

Borlaug made a weary journey homeward to Mexico. Through the droning hours his mind roamed over the implications of the developments in Pakistan. If the new president complied with the economists' arguments and broke faith with tens of millions of rural people, the wheat manipulators would find it easy to corner the market and create a false shortage. The food division between the east and west wings of the Pakistani nation was clear. A further shortage of grain would add to the anger of the disgruntled people of East Pakistan. West Pakistan's terrain, the 1,000-mile-long Indus Valley, with extensive irrigation and canal systems, was a fine environment for the dwarf wheats and the new dwarf rice strain. Both cereals were grown there. East Pakistan, with extensive delta country from the Ganges and the great Brahmaputra outflows, had a rainfall averaging a hundred inches a year. It was a wet land, primarily suitable for rice, and it was ridden with virus disease and blight. Chandler's new rice strain, IR8—which was doing so well in the rest of the subcontinent—had suffered badly in East Pakistan. But newer disease-resistant types from IRRI were beginning to show promise but had not yet been multiplied for commercial use.

Before the seed harvesting was completed at CIANO late in April, news from Narváez and Hanson confirmed Borlaug's fears. The economic ax to which President Yahya Khan gave his approval chopped 25 percent from the guaranteed price of wheat.

Four months later bitter disappointment turned to indignation. Borlaug and Narváez were summoned by the president to answer allegations that they had oversold the promise of the wheat harvest to Pakistan. Once more Borlaug traveled across the world to face a crisis. Narváez, he knew, could make an extremely effective reply to the charges. But Borlaug was the chief architect of Pakistan's wheat revolution and his honor was at stake. Narváez met him in Karachi, and together they went to Lahore to face the accusations. On the way Borlaug became ill. He was sneezing and had a sore throat as they checked into the hotel.

Yahya Khan was distant and frigid. He made his disapproval of them clear. They had much to answer for in Pakistan's current crisis, he said. They had overestimated the wheat crop of the 1969 rabi season; they had allowed enthusiasm to run away with good judgment. Food prices were soaring. Because of their forecasts, orders had not been placed for overseas supplies of grain; and there was now a shortage of food in the country. They were to blame and he demanded an explanation.

Borlaug was still feeling ill. Waves of fever ran through him. He said, quietly, simply: "Dr. Narváez and I will go out there, into the country, and look and see what the situation is, Mr. President. When we have our facts, we shall come back here and answer you—personally."

They went back to the hotel. He felt faint with the sweat and shiver of influenza. He went to bed and took his temperature. The thermometer registered 104 degrees but he felt chilled inside. He said to Narváez: "Pakistan is in a hell of a state, Nacho. And so am I! Give me a day or two, and we'll get under way."

Two days later he climbed, shaking, from his bed and

they left. They went up and down the country for two weeks, persistent and ruthless in their investigation. They asked, cajoled, argued, and pressured. They came back with their evidence and sent for Anderson and Aresvik. When they sat down together in the hotel, Borlaug was reminded of the night with Hanson when they wrote the first memorandom for Ayub Khan—four men condensing a hard-hitting report. Only this time it was for a different president.

All four authors went together to see Yahya Khan. Borlaug was brutally direct. He told Yayha Khan that the report "laid the truth on the line"; it showed that the government had played into the hands of speculators. "There is no food shortage in this country! There are plenty of unprincipled grain dealers—and they sure have the government over a barrel—but there is no food shortage. The harvest was about as expected, and the yield was not overestimated. There is nothing that could have been avoided. If there is little for people to eat, the government created that situation. You signed the order that has brought this trouble. You broke faith with your own people. You took away what had been promised them. We are not to blame; the farmers are not to blame. The government is to blame! In this office you signed the document that opened the way to hoarding and crooked speculation. And I will tell you to your face, sir, that speculation and manipulation are rabid in this country right now."

Yahya Khan was irate. He shouted at Borlaug and his companions. "Why do you blame me? They came here, into this office, to persuade me—the expert planners. If I am misled, it is they who have misled me."

Borlaug said: "Mr. President, that is possibly the only thing about which we will agree."

Aresvik tried to speak, but Yahya Khan waved him to silence. He said to Borlaug: "Well—what can I do now to make things right again?"

Borlaug said there was one solution: recapture the faith of the farmers and give them new trust. "And how do I do that?" Yahya Khan asked. Borlaug considered. Six weeks—eight at the very most—remained before the beginning of the next wheat planting season. There was very little fertilizer and not much chance of an emergency shipment from the United States. The shipping strike there was still in progress. "Things are screwed up for you, Mr. President. You have got to clean out your coffers and buy fertilizer quickly, wherever you can get it fast. You have to announce a higher guaranteed wheat price for the coming season—and you have to stick to that! You have to find money for credit. Perhaps, if you put the wheat price up a bit higher than it was last year, you might have some sort of chance of getting your country out of this mess."

Yahya Khan asked at what level he should fix the price of wheat.

Borlaug shrugged his shoulders. "I can only advise you on agricultural policy. You are the leader, the politician with economic advisers. I suggest you ask them. Ask them how to get that fertilizer here, how much to pay. Then ask how much they saved by disillusioning the farmers earlier this year."

Yahya Khan was distressed. "I took the advice of those counselors," he said. "Then it is their fault."

Borlaug said: "Yes, Mr. President." They left the office.

Back at the hotel he said to Aresvik and Narváez: "He is a different breed of man from Ayub. He should have stuck to soldiering. It might have been better for him—and for Pakistan."

Yet Yahya moved to acquire fertilizer and issue wheat price guarantees. The surge returned to wheat production, as was predicted. At the end of the rabi harvest, in May 1970, a keen Pakistani correspondent wrote to his news agency in Karachi: "The Green Revolution has again caught up with the illiterate farmer of West Pakistan. In one quick sweep it has transformed his meager holdings into rich golden wheat fields. His methods are those used by his forefathers along the Indus Valley 2,500 years ago—but the new seed from Mexico, intense instruction, and mushrooming fertilizer supplies have pushed him into the breakthrough."

One after another new philanthropic agencies, public institutions, banking groups, and countries were becoming associated with CIMMYT. New association meant new funding. Borlaug was deep into the sponsoring of programs in Africa and South America. The forces under Wellhausen, as general director, expanded in the scientific institutions in Mexico, particularly at the new Chapingo center of El Batán, at Toluca, and the CIANO and Poza Rica stations. These formed a central axis of operations supervised by a globally selected board of directors—men from Asia, South America, the Middle East, and Southeast Asia.

In this growing work force the issue of higher nutrition had been accorded priority, second only to the burning need for grain. In the poverty-stricken Caúca Valley, near Cali in Colombia, the Opaque 2 corn discovery had been pressed into use in human therapy by 1970. The Rockefeller Foundation had sponsored work that applied the Opaque 2 preparation to undernourished domestic animals —with striking results. Now Dr. Albert Pradilla was at work at the University of Valle Hospital treating children who were chronically ill from protein starvation. He was devising methods of introducing vitamin additives to the hybrid cornflour and feeding it to starving children who had been taken to the hospital to die. By 1970 Pradilla had amassed some remarkable case histories. In a Rockefeller report he wrote: "There is significance in the fact that we have a potential source of low-cost, high-quality protein that can do much toward preventing malnutrition. This means life or death for millions of children. It is obviously better to prevent malnutrition than to treat it."

The diversification of activity at CIMMYT gave impetus to Borlaug's work. In 1970 Mexico's average yield of wheat grain per acre was 1.5 tons—against 825 pounds in 1945. Approach and inspiration were staunchly emphasized in his work with each new group of wheat apostles that came to Mexico. Borlaug maintained that the real apostles were the missionaries who went into the world's fields to see that food supplies were insured. In October 1970 he worked with a large group of such missionaries—men from all parts of the world.

In all this work—the crossbreeding, the constant searching for better two-gene or three-gene dwarf wheats—the triticale flame burned brighter and brighter.

In the fall of 1970 the triticale plots at Toluca were verdant, vigorous—and stimulating to their creators. A new line called Armadillo, developed by Zillinsky, had been identified by Eva Villegas and her team of biochemists as showing high promise as a weapon against malnutrition. The plots of triticale seemed unusually heavy and thick; they had bountiful foliage, shorter straws than the first tall varieties, and heavy heads. They looked similar to their parents but were neither wheat nor rye. They were something in between. They had the long rye rachis, the seed-bearing stem—longer than a man's hand, filled with plump kernels. The close-packed plants grew exceptionally green with a glossy texture and showed resistance to normal rust attacks. Zillinsky had crossed most of the lineal defenses of the dwarf wheats into the plants and had preserved the photoinsensitivity. Delighted with the progress, he and Borlaug sent sample seeds to colleagues for testing at co-operative stations in Ethiopia, Spain, Argentina, and India, as well as in several other Asian countries. Yield tests were also made at Toluca, in the new experimental plots at El Batán near Chapingo, and at selected locations in the southern United States. From one country after another came encouraging reports. From Ethiopia Dr. F. F. Pinto wrote: "The triticale have performed as well as the best bread wheats and are superior to all others." At Toluca the Armadillo strain gave a yield of 2,800 pounds per acre—against 2,600 pounds for the best-yielding dwarf wheat, Pitic.

The nutritional tests were also heartening. Borlaug was told that Dr. Fred Elliott of Michigan State University had found that the different strains ranged from poor to excellent, with the best giving a nutritional value approaching that of egg protein. At the time Borlaug and his team had close to 200 different lines of triticale, but increasingly they realized that they were working from a narrow genetic base.

As with many scientific advances, the triticale brought as many problems as promises. New genetic combinations in the germ plasm held latent traits that had never been given expression in the parent wheats or ryes. These imposed a severe and rigid selection process on the workers. One gene apparently released an enzyme which produced grain shrivelling. Kernels rapidly grew plump and fat—until nearly ripe, when the enzyme appeared to break down the stored starch. The grain dried and became wrinkled and unattractive. Other types developed weak necks and others tended to lodge soon after fertilization. Some showed susceptibility to diseases afflicting rye, some to those afflicting wheat. But one by one the faults were weaned out and the program was expanded through selection of the better performers.

The researchers also encountered the same problem that had assailed the Opaque 2 corn types: the triticale kernel tended to make a soft gluten interior, which affected its prospects as a good breadmaker. The team struggled with these perplexities, and for Borlaug and Zillinsky the task was fascinating with its high potential. There was a long way to go before the plant would have real commercial value, but seeds of the most promising lines had "escaped" from the stations in Mexico and elsewhere.

Borlaug said at that time: "There is no question that the triticale are being sold as a commercial crop too soon. It is premature to be pushing this plant the way it is being offered, as a supergrain. Much more work needs to be done, though I must say it is very promising indeed." As with the series of dwarf wheats, all that could be done was to press on with the triticale work at top speed until truly superior varieties could be multiplied for sound commercial production.

In a discussion at that time Zillinsky said to him: "I'm certain, Norman, we shall solve that last headache of the shrinking grain. I *know* we can give the world something extra in nutritious grain. We'll get over the hump for sure, by the mid-1970s."

Borlaug replied: "Once we get triticale over that hump, Frank, there is fantastic change looming. We can try to cross all sorts of plants that have never been successfully crossed. My mind still asks, as it did many years ago in this valley: Why can't we cross wheat with rice? Of all the cereals this intrigues me most."

Triticale, the new breakthrough, the promising supergrain, was already a step forward, a move toward new advances that would demand new organization and new men. "A few people are going to look at us again as though we are mad, Frank. But I think it is all possible and worth a try at the right time. There is change on the way."

In mid-October 1970 he worked in the fields at Toluca with the latest group of young scientists from around the world. Borlaug was in his fifty-seventh year, and he worked with them, spending hours with his back bent, crouched over plants in the tough initiation of the harvesting season. During the physical stamina trials, he gave his usual talk to twenty-five young wheat apostles from twenty different lands.

"We seek higher yields, but we must drive on for nutritional potential also. We must face the fact that there is a limit to available land. If we increase yields, we make less demands for land area and we reduce costly inputs necessary to opening new areas. We grow more food—and better food—but there is a dreadful, steady ticking in our ears, the ticking of the population count! Each year when I see a new gathering of faces at this school there are another 75 million mouths to be fed in this world. So when you think the work is hard in this school, think on those facts and what you have to do to defeat them. You work to give the world time to come to its senses and control the population bomb."

Sixteen

It was normally quiet in the Borlaug house, but on the morning of October 20, 1970, it was quieter than usual. Mexico was mourning for its former president, Lázaro Cárdenas, the honest, beloved leader who had exiled Plutarco Calles, implemented the ejido system, and founded the Mexican Federation of Workers. The city, the whole country, was standing still.

The silence in the Borlaug home was shattered at a little past seven with the ringing of the telephone. Margaret was awake, dressing, but the sudden ringing was jarring anyway. She answered: the call was from Norway. Margaret accepted that information in stride; she was accustomed to receiving calls for her husband from places far more exotic than Scandinavia. A man's voice replaced the operator's; his Norwegian accent reminded her of Norman's father. The man identified himself as a reporter for the Oslo *Post;* he wanted to speak to Dr. Borlaug. Margaret told him that Dr. Borlaug had left an hour earlier and could not be reached by telephone. The reporter was disappointed but explained the reason for his call. In a few hours, he said, Norman Borlaug would be announced the winner of the Nobel prize for peace.

Margaret was incredulous. Maybe this was a hoax. "Where did you say you were calling from?" she asked. The reporter told her. How, Margaret wanted to know, did he know before anyone else what the Nobel Foundation had decided? This was her first word of it, and a newspaper was not an official source. "It will be official all right," the reporter said, and laughed, explaining that newspapers had ways of finding out such things. "It will be announced officially here at five in the afternoon, Oslo time." Margaret thanked the reporter and hung up. The telephone rang again: another caller with the same information. It rang again, and again, and again.*

Margaret wanted to be the first to tell Norman the news. Before the phone could ring again, she called Dr. Robert Osler, the senior administrative officer at CIMMYT in Mexico City, who agreed to send a car to her house to drive her into the mountains. Osler also sent his wife, Elane, to ride along to keep Margaret company.

After she spoke to Osler, Margaret telephoned Ed Wellhausen at home. Wellhausen was still general director of CIMMYT, but he was ill in bed, and she wanted him to know. She spoke to him briefly, and when she was through the car was there. As she went out the door, the phone started ringing again. She let it ring.

It was nearly ten o'clock when the car bumped down the rutted road along the edge of the Toluca station wheat plots. Norman Borlaug was working with a group of trainees. On one side of him were two young Rumanians; on the other a Brazilian. He saw the car approaching and was curious. When it stopped and Margaret alighted, Bor-

* Contrary to the popular notion, the Nobel awards are not a purely Swedish enterprise. The nominations are made by the Norwegian parliament and sent to the Swedish parliament, which makes the selections—except for the Nobel peace prize, which is determined entirely by the Norwegians.

laug had a sudden chill of apprehension. He noted that Elane Osler was with her. Something had happened: his parents were ailing in the Cresco home; Jeanie or Billy had suffered an accident. He dropped the wheat plants he was holding and ran between the rows, his feet thudding on the earth. He was panting when he reached her. "What's up, Margaret?" he asked. "What's wrong?" But she was smiling. She held out her hands to him.

"Nothing is wrong, Norman, nothing. You've won the Nobel peace prize. That's all."

His expression was unchanged. They stood together, holding hands in the Mexican mountains, near clumps of his wheat. Margaret could see that he did not want to believe the news. Finally his expression changed. "Who says so?" he asked in disbelief. "Who told you?"

She explained about the call from Oslo and the calls that had followed. He seemed apprehensive. Margaret watched his lips pucker, the lines deepen above his eyes. "It doesn't sound very official to me, Margaret," he said at last.

But so many people were calling, she said. They wanted to talk to him. She wanted him with her. She didn't want to handle everything, all the inquiries, herself. "Come back to town with me, Norman," she pleaded. "It might be best."

Borlaug looked from side to side like a trapped man. The trainees had stopped working and had gathered together, watching, sensing a crisis. He saw the young CIMMYT geneticist, a wheat plant in his hand, standing with the others.

"It could be speculation, couldn't it?" he said to Margaret. He kissed her, smiled at Elane Osler, and added: "Best thing is for you to get back to town. I'll work on here. We'll see what happens."

Margaret knew that further talk was useless; he had made his decision. She pleaded no more as he walked with her and Elane Osler back to the car. She watched him return to the trainees, and as the car rolled down the road she could see him back among the wheats, his head down. Afterward she wondered if he had had some inkling, some premonition of the assault about to be launched on their lives, of the demands, arguments, controversies, and criticism that would burden him and separate them far more than the wheat work had ever done.

Standing among the rows of wheat, after watching Margaret's car disappear on the road back to Mexico City, Norman Borlaug held on to his last moments as a private man. He told the others about it, emphasizing that it was only a rumor. He reminded them of the work they still had to do, rumor or not. Within a week he was due in Argentina and Brazil to advise on wheat-growing acceleration in those countries.

His last moments of privacy lasted only forty minutes. Then, as he said later, the Nobel peace prize "struck my life like a typhoon."

It began with the arrival of an American television crew in the field. "Which one of you guys is Borlaug?" one technician shouted. Identifying himself, Borlaug said that since he had not yet been informed officially of the prize, he didn't see how he could make a statement. "You had just better be prepared to believe it, brother," the television man said. "It's on the world's newswires. We don't make that kind of boo-boo."

Other crews arrived. They began arguing among themselves over priority in the right to interview him. Borlaug felt a queasiness deepening in his stomach. It would be worse than this in Mexico City. He was glad he hadn't gone back with Margaret. Other media people reached the field. Across the road from where he and Dick Spurlock had first cut the Supremo in 1947 the eye of world publicity focused on him.

Two hours later Robert Osler came racing onto the

wheat field. The maelstrom had enveloped him too. There was no way to avoid it he told Borlaug: the demands for information, for interviews, must be met. Osler had called an international press conference for Borlaug at five o'clock, with television hookups to follow. Borlaug realized that he had no choice and agreed to return with Osler. Just then one of the cameramen trampled the end of a line of wheats. Borlaug cursed him.

He wanted to go home, change his clothes, and pick up Margaret, but when he neared the house he saw it was under siege by a legion of men with electronic equipment. Driving past without stopping, Borlaug and Osler headed for Ed Wellhausen's home nearby.

Wellhausen had been stricken with a pulmonary embolism; he was wan and thin and trying not very successfully to keep himself calm in the excitement. "I suppose it's true, Ed," Borlaug said as the two men shook hands. "If it is, it's for all of us, you and me, and Dutch Harrar—for the whole darned army. They've just singled me out to say thank you."

"This is a triumph, old friend," replied Wellhausen, "for us and all our brother scientists. You must be the first agricultural scientist to get the *peace* prize. The world has been served notice of the link between food and peace."

When Borlaug faced the newsman, he still wore his muddy boots and held his baseball cap in his hands—until Osler took it from him. The first session lasted two hours. When a well-read journalist recalled the smear from Pakistan and India—the assertion that the new wheats would make people and animals impotent—Borlaug snapped into the microphone: "That would be one hell of a thing if it were true! This Green Revolution's a moving thing; it still goes forward. But it gives the world no more than twenty or thirty years' breathing space to do something about the real problem—the population explosion! The dwarf wheats won't make utopia for you. They can't make clouds in a blue sky, and they can't stop people from procreating. If they could, then they should lump all the Nobel prizes—peace, chemistry, medicine, what have you—into one award and give it to CIMMYT, because the greatest threat to man in this world, the biggest cause of hunger, and the most terrible menace to peace is this multiheaded population monster."

While Borlaug taped a series of television interviews, Margaret and the CIMMYT wives organized a staff party. It was nearly three in the morning when the Borlaugs got home, and the telephone was still ringing. Norman took the receiver off the cradle. As they fell exhausted into bed, Margaret remembered to tell him that Billy and Jeanie had managed to get through; so had Dr. Stakman. Margaret had more trouble getting to sleep than her husband did. It had all been so exciting—not just the prize today but all the years leading up to it. She had been lonely, times had been difficult, but his career had been nonetheless exciting. Now what? You could take the telephone off the hook to get some sleep, but you couldn't turn off the world, or turn away from new obligations and responsibilities. Just before she fell asleep Margaret felt a chill of apprehension, though she didn't know what exactly she was anticipating.

It was business as usual the next day. In the morning Borlaug drove up to Toluca to finish the harvest selections. But he found no peace in the mountains. Reporters swarmed everywhere.

At CIMMYT headquarters calls, letters, telegrams, and requests for television interviews to be beamed to Europe via satellite kept the staff busy. Many of the letters could wait, but some called for immediate response. One came from Mrs. Gandhi, another from Yahya Khan, others from prime ministers and presidents. Richard Nixon sent a cable saying Borlaug had given Americans reason to feel proud. There were letters from Henry Ford, from the Rockefeller family, from universities and governments. Harrar, now president of the Rockefeller Foundation, telephoned and spoke to Borlaug; then he sent Dr. Dorothy Parker, the first librarian at the old Office, to help the besieged scientist with his correspondence. Borlaug decided that he must cancel his tour of Brazil; he had too much to do now, and the Nobel lecture to prepare.

He couldn't work in Mexico City; there were too many interruptions. So he hastily packed a bag and flew to New York, where, he told Margaret, he thought he might be able to "get lost among the crowds." His anonymity lasted two days; then newsmen tracked him down. He fled to Iowa to see his parents. Terribly crippled by arthritis but alert, his mother was thrilled. But the news had come six months too late for his father. He was almost completely blind and was losing his mental powers. "Somethin's goin' on," he said. "I know. I can't make out what it is." Norman Borlaug wept as he held his father's hand.

The Rockefeller Foundation summoned him back to New York. The FAO in Rome was applying pressure on the foundation to convince Borlaug to speak at the annual FAO conference in November, to be opened by Pope Paul. Borlaug did not want to do it. "I know well where the pope stands on population, and it is not where I stand. It would be dynamite for me to speak from that platform."

He returned to Mexico City. The FAO continued to bring pressure to bear on the Rockefeller Foundation. Borlaug pleaded sickness and within a week actually went to bed with a severe attack of influenza. As he lay recovering, he arranged to take his whole family—Margaret, Billy, Jeanie and her children, and his two sisters—to Oslo for the presentation. His pleasure was dimmed, of course, by the fact that he could not take his parents. He was flattered to learn that Roberto and Teresa Maurer were going to go along; they had cashed in some insurance to pay the fare. "How could we miss such a thing?" Teresa said to Margaret. "He has changed our lives and is our beloved friend." Other Mexican farmers announced their intention of joining the contingent to represent the rural peoples of the country where the experiment had begun.

It was, then, a formidable entourage that reached Norway. Borlaug was immediately summoned to a private audience with King Olaf at the palace. The monarch told the scientist that he was familiar with the region where Borlaug had been born and raised. Twice during World War II Olaf had visited Spillville, where Dvořák had worked.

Norman Borlaug received the Nobel prize for peace the next day, December 10, 1970—the seventy-fourth anniversay of the death of Alfred Nobel. (The other Nobel prizes were awarded simultaneously in Stockholm.) Madame Aase Lionaes, chairman of the peace prize committee, spoke of Norman Borlaug as an "indomitable man who fought rust and red tape . . . who, more than any other single man of our age, has provided bread for the hungry world, . . . [and] who has changed our perspective." In his acceptance speech Borlaug said that he did not see the prize as an award for an individual scientist: it belonged to an army of hunger fighters across the world who were still fighting. "The Green Revolution has not yet won the battle for food, because mankind is breeding too fast."

After collecting his parchment and medallion and a tax-free check for $78,000, Borlaug and his family and friends attended the ceremonial dinner and ball. When the party ended, he and Margaret drove upcountry to visit the birthplace of his parents and enjoyed a few days of smorgasbord, singing children, and waving flags—and the ringing of bells of welcome across the snow-covered country. In the several speeches that he made he repeated the theme of his work, a theme that was now linked, thanks to the Nobel prize, with world peace. For if the fight against hunger was successful, it would bring peace. Empty bellies, he said again and again, were the drums of war.

Borlaug found walking in the homeland of his ancestors and talking with the Norwegians deeply satisfying. But he had too few days to spare, and he and Margaret soon had to fly back to the United States.

The snow in Cresco was deeper than they had seen it in Norway. Huddled on the wide prairie, the town of 4,000 made a picture-postcard Christmas scene, with colored lights and glowing windows. A snow-rutted road ran out toward Saude and the old one-room wooden schoolhouse from which, half a century ago, he had struggled home in deep snow to the fragrance of his grandmother's baking bread. He rented a car and drove Margaret out to look at the old farms. He said he would like to use part of the Nobel money to buy his father's farm back for the family. If he could make the deal, he remarked, it would be the first property he had ever owned.

The outside world touched Cresco on December 20, 1970, which the town fathers had designated Norman Borlaug Day. In the high school gymnasium Borlaug shared the platform with many. Iowa's Governor Ray presented the scientist with Iowa's outstanding service award. Dr. Dave Bartelma was back for the occasion, and at his side, down from Minnesota, was perhaps the proudest man of all, Dr. E. C. Stakman. Officers from the Rockefeller and Ford foundations were there, along with representatives from a variety of institutions, the U.S. government, and universities. His father was too sick to come, but his mother was present in her wheelchair, surrounded by his sisters and cousins. And the townspeople had turned out to honor their famous son

—but, Borlaug noted with regret, not many *young* townspeople.

He described briefly for the gathering the background of the Green Revolution and then expressed his concern at what he called the "drift among youth toward complacency, aimlessness, and indifference to the suffering of their fellow men." In a world threatened by widespread hunger the young were backing off, opting out from responsibility. He did not specify where the blame lay, but he thought that doomsayers were instilling in the young a sense of hopelessness. And many of the pessimists, he knew, formed the most extreme and conspicuous element of the ecology movement.

Borlaug became the declared enemy of ecological extremism in the early weeks of January 1971. He sat one day with Haldore Hanson—then the designated successor to Wellhausen as general director of CIMMYT—after reading an attack in an American newspaper on the menace to the environment and wildlife posed by the use of agricultural chemicals: fertilizers, weedicides, and pesticides such as DDT. "Damn it, Hal," he said. "I first handled these things thirty years ago. I've lived and worked to the armpits with this sort of stuff. What in heaven's name are they writing these exaggerated stories for? There's not one case in world medical literature that tells of harm to human beings from DDT when it's used properly. It's all moonshine cooked up by people who don't realize the damage they're doing." If necessary, he would fight the frenzied fringe of the environmental movement to the grave rather than let it force decisions in the United States that would create misery among the underfed peoples of the world.

Hanson tried to calm him. He saw Borlaug's anger growing every day but told him not to fight a lone battle in the ecological arena—not at this stage of his career. Borlaug disagreed. This was a new, challenging campaign confronting him. He himself had raged against population growth and warned of its dire consequences, but he had never believed the extravagant claims of some ecologists who predicted an imminent death to a depleted planet. Borlaug wanted to guide public opinion back to a sensible middle ground. There were terribly serious problems to be dealt with—but they could be dealt with, and not only with extreme measures. The Nobel prize, he told Hanson, had placed an additional obligation on him, not freed him to become an elder statesman, above all battles. He had to use his new distinction where it was needed.

"I knew full well I was stepping into the eye of a hurricane," he recalled later. "I faced critically important food production tasks in Asia, South America, Africa, and the Middle East. I knew that what happened and what was decided in the United States had impact and bearing on those areas. No country is an island any more. I had to speak out to say there were two sides to the environmental issue. I was at once drawn into the vortex. As soon as you start to challenge an emotive issue you get attacked by feverish, committed people. I was subjected to insult, mudslinging, trashy verbal assault. I knew all that would come. But what could I do but accept the responsibility?"

The era of ecological awareness had been launched by the sober book *The Silent Spring*. After its publication in 1962, a succession of books—many by academics, many anything but sober—had appeared, asserting that modern man was killing his planet. To Borlaug, not only were these books one-sided and unrealistic, but their pessimism was damaging to youthful minds because they sought to correct abuses by means of fear which, in turn, prompted overreaction.

Borlaug knew that what he had to do would bring him vilification. He would be attacked; his work would be demeaned by people who could not do what he had done. And so at fifty-seven the Nobel laureate who was called the father of the Green Revolution had another battle on his hands.

He was not so alone as Hanson had predicted he would be. Other reasonable men were seeking to bring moderation into the ecology movement. Dr. John Carew of Michigan State University debunked the "balance with nature" theory in verse and painted a picture of a world where laws prevented the use of fertilizers and pesticides—a world where insects, weeds, and diseases flourished and whole communities starved, a world torn by war for remaining food resources. It was a world in which, eventually, people were back where they had started, living like animals in the wild—but in "balance with nature." Carew wrote to Borlaug that he had been berated for the work, and that the criticism revealed "how intelligent people saw scientific agriculture as a force in opposition to environmental quality. . . . Fortunately, I can see signs that some people are becoming more rational in their thinking. It helps when Dr. Handler of the National Academy of Sciences speaks out against environmental alarmists; when the editor of *Science* writes in recognition of the dependence of our food supplies on fertilizer and pest control. Somehow or other, I can forgive the public for having been misguided, but I reserve my anger for the previously 'pure' biologists, botanists, and others who suddenly have all the answers to our pollution and social problems."

Borlaug was not completely opposed to the tenets of the "balance with nature" advocates. He was not enamored of chemical fertilizers and insecticides for their own sake. In his Nobel oration he had spoken of eventually establishing a symbiotic relationship between wheat and the bacterial colonies that fixed nitrogen in plants—in other words, of breeding plants that manufactured their own nitrogen nutrition. He had also envisioned a day when the breeding cycles of insects could be controlled by biology, rendering pesticides unnecessary. But in 1971 self-fertilizing plants and harmless insects were distant dreams; in 1971 the world still needed fertilizers and pesticides. If insects were uninhibited by controls, they would destroy a large part of the grain foods in the United States and would wreak even greater havoc on the crops in the semitropical lands, across South America, in crowded India, in Africa. Winters in temperate climates do much to check the buildup of pests. But this is not the case in true semi-tropics and tropics—where pests must be controlled lest millions of humans perish.

As the weeks went by, Borlaug kept to a strenuous schedule of speeches, meetings, and field work. He attended symposia and seminars and conferences; he addressed students and industry; and always his words stung the extremists of the ecology movement. At a farmers' conference in Minneapolis he was booed by student followers of the organic farming movement when he said that plants could not tell the difference between nitrogen produced in a factory and nitrogen bred by bacteria in the "natural" fertilizer called humus, made of decomposing organic waste. Outside the hall in the Minnesota metropolis he was surrounded by placard-bearing youths. Borlaug stopped to talk with them, informing them that the chemical fertilizers that they so despised stood between health and hunger for millions of underfed people. "I too don't like the idea of silent springs," he said. "But I refuse to be an alarmist. I will not bark up the wrong tree and I will not be forced into quitting on the aids we use to produce more food for man."

During another lecture, Borlaug was pressured by students who advocated the return to completely organic farming so that "man could live in balance with nature."

"Okay, let us pursue this philosophical argument to its conclusion," replied Borlaug. "If you are going to live in balance with nature, as you put it, you must be prepared to go all the way with it. To achieve this balance you are going to have to do without a lot of things: fertilizers, pesticides, preservatives, vaccinations, prescriptions, medical care. The conclusion of your hypothesis is that you will have to cope with disease because your intestinal parasites —the roundworms, the tapeworms, whatever infection you might have—all these will have as much right to exist as you do. I am certain that if we were to live as you desire that the world's population would drop precipitously. But think of the human suffering and misery, the social and political chaos this would cause. Would you modern utopians really want to live in a world like that? I doubt it. So why don't you try to solve the world's problems by using common sense."

In the summer months of 1971 Borlaug undertook inspection tours of the fields of Latin America, North Africa, the Middle East, and the Indian subcontinent. Everywhere he went he emphasized that the battle for food had not yet been won. He repeated the words he had used in Oslo: "Tides have a way of flowing and then ebbing. We may be at high tide now, but the ebb tide could soon set in if we relax our efforts and become complacent. We deal with opposing forces: the scientific power of food production and the biological power of human reproduction."

In India he received more honors and awards. At the Uttar Pradesh University in Pantnagar he was conferred with the degree of doctor of science *honoris causa* in company with other recipients, including his Indian colleague Dr. B. P. Pal. The citation recognized Borlaug's "great contribution to the wheat crops in particular, and other cereals in general, and to initiating the Green Revolution in the world."

Once again he was summoned to the New Delhi office of the prime minister to review India's food production problems. He was a great man now, and his word carried weight. He knew it, took advantage of it, and reminded Mrs. Gandhi of the growing need for fertilizer supplies, for expansion of the breeding work being done by the agronomists.

Touring the poorest regions of India, he saw that the margin between sufficiency and disaster was still awfully slim. Like all the nations of the underfed world, India needed grain reserves to which it could turn in emergencies without meeting inflated world market prices. Borlaug advocated the idea of strategic world granaries that would combine national stocks with international reserves. In future speeches he would have to remind people of the lesson of the story of Joseph in the Bible. The years of plenty would have to provide enough for the years of want.

Two months after the ceremony in India Borlaug was called to Washington to testify at a hearing of a subcommittee of the House Committee on Appropriations. "Now everyone talks about the environment," Borlaug told the congressmen. "In universities I am told, 'But you are one of the people contributing more than anyone else to the deterioration of the environment by use of heavy doses of chemical fertilizer to produce more food.' That is a great oversimplification of fact. The fertilizer used in the developing nations is what is economically feasible. It is much less than that used in Western Europe, the U.S.A., or Japan. In other words, in the developing nations you must put into the soil and extract out of the plant a very high percentage of that nutrient. Little, if any, goes into the streams. In the first place there aren't many streams. Water is a most limiting factor, and we are confronted in trying to lift food production with oversimplified attacks. Everyone wants a simple answer to a very complex question."

During the year Borlaug learned that environmental groups were preparing a campaign aimed at banning the use of DDT. He was aware of the great contributions made by this insecticide that had brought malaria under control for the first time in many semitropical and tropical countries. What would happen if DDT was no longer used and malaria returned? He had been asked to deliver the prestigious MacDougall memorial lecture, which was given each year at the November conference of the FAO in Rome. He had been unable to attend the previous year. But now he needed a world stage to fight the forces assaulting fertilizers and pesticides, and he accepted the invitation.

Borlaug discussed his simple strategy with Dr. Stakman. "This whole movement is too lopsided," he said, "too distorted in the face of the misery of hunger in the world. No one loves wildlife more than I do—no one. But I also love man and I want to see his children fed. I am going to Rome to speak aggressively on this eco-fever issue. There will be no holds barred. The gloves are off."

He titled his lecture "Mankind and Civilization at Another Crossroad."

On the first Friday of November 1971 Borlaug arrived at Kennedy Airport in New York to board an evening jet to Rome. As soon as he entered the terminal, he heard his name called on the public address system. At the airline desk he was told that an urgent call was waiting for him in the traffic office. He picked up the telephone there and heard the voice of his sister Charlotte. "Norm?" she said. "I'm terribly sorry; I had to reach you before you left. I'm sorry. Dad died today."

It was not unexpected, but for a moment Borlaug was speechless. His quiet, sweet father had lingered in his half-world, no light penetrating his sightless eyes, little penetrating his dying mind. In the moment's silence after he heard the news Borlaug recalled how they had talked on the telephone when his grandfather had died. He had not been able to attend old Nels' funeral because of college exams. He had not even been able to attend the funeral of his little son Scotty, bcause of the distance and lack of money. He did not want to miss this one. He asked when was the latest date they could have the funeral, and Charlotte said they could wait no longer than Tuesday. Borlaug said he would try to be there.

After he hung up, he put in an urgent call to the office of the general director of the FAO in Rome. When he reached a senior official, he explained his problem. "I know this thing has been set up for months," he said. "However, I will not miss my father's funeral. If you can advance the date of the lecture to Monday and make it early enough for me to take a jet back to New York that same day, I'll keep the appointment." He was assured that the lecture would be scheduled immediately after the formal opening ceremony.

The media were alert to his attack. The biggest array of newsmen, cameras, and tape recorders he had ever seen confronted him in the FAO conference hall. He took a deep breath, reminded himself to control his tone, and, speaking in a measured voice, began.

"Peace cannot be built on empty stomachs." Those were the words of Lord Boyd Orr, the founder of the FAO, and Borlaug included them early in his speech. Today, he said, there were two worlds. The affluent, developed nations formed a quarter of the whole of mankind—living in luxury never before experienced by man. But three-quarters of the human race formed the forgotten world where poverty, hunger, and starvation prevailed. The Green Revolution had helped temporarily to alleviate some of the hunger, but its progress was modest in comparison with world needs. He called for the establishment of a network of strategically situated international granaries on which nations facing famine could draw. "We will be guilty of criminal negligence without extenuation if we permit future famines. Humanity can no longer tolerate that guilt." World shortage of grain could be met by a 30 percent increase in crops from the major grain-producing nations, but that would not cure the problem because nations with weak economies could not bear the expense of importing grain. They had to grow enough to feed themselves. The Green Revolution had been a step toward this, despite the criticisms of some Western commentators.

The success that had been won, Borlaug explained, was temporary. "It has given man a breathing spell. Implemented fully, it can provide food during the next two or three decades—hopefully, a period for man to balance population growth to a decent standard of living for all born into the world."

Next Borlaug attacked the extremists of the environmental movement. "Deny agronomists the use of fertilizers and other chemical aids and the world will be doomed—not from poisoning, as they say, but from starvation." A powerful group of extremists, he said, was provoking fear, and they were busiest in the most advanced, best-fed countries. Their preaching was especially well received by affluent professional people, cityfolk who tended to regard countryfolk—people who grew the food they ate—as bumpkins or dullards. Now these urban dwellers and their crusading were actually, if inadvertently, threatening the survival of nations on the other side of the earth.

Borlaug also warned that success in banning the use of DDT in the United States would almost certainly lead to a movement for a worldwide ban. This ambition was not sensible—especially in view of the fact that American cities continued to dump billions of gallons of raw sewage into lakes and rivers. The doomsayers blamed the ills of society on science and technology but never acknowledged the state the world would be in today if it did not have the food provided by the very science and technology that they were trying to discredit. "For less privileged city dwellers, pollution does not mean fertilizing chemicals or DDT; it means slums infested with cockroaches, rats, and mice, rotting garbage, lack of fresh air, and a crying need for health services." Despite this great human need, environmentalists in and out of government, partially informed people in the communications media, and others had involved their energies in a crusade designed to end the use of agricultural chemicals. He told the FAO conference: "They give no thought to the end result of such action—the eventual starvation and the resultant political chaos sure to engulf the world. Why is such a short-sighted policy pursued?"

When he left the platform, the ranks of newsmen rushed to surround him. Officials of the FAO shielded him, pushing the reporters back and telling them he had suffered a personal loss and had no time to answer questions. Later the FAO and WHO supported his arguments on the use of DDT in agriculture and public health programs. "Until a cheap, safe, and efficient substitute is produced and made easily available," the FAO statement said, "there is no alternative to the judicious use of DDT, especially in the developing world, to increase agricultural productivity, to protect the health of and to feed the growing number of people on our planet."

Borlaug reached Cresco two hours before his father's funeral. The burial was on the grounds of the Lutheran church in Saude, in the plot allotted to the Borlaug family.

When the service was over, Borlaug paused for a moment at his grandfather's headstone, and the words he had heard as a small boy came back to him: "Common sense, Norm, that's what the world needs. Education and common sense."

How odd, Borlaug thought, that his grandfather's philosophy was needed today more than ever before.

Seventeen

By the end of January 1974, the disaster Borlaug had spoken about earlier seemed to be materializing around the world. Newspapers that had overlooked his earlier statements now featured his words on their front pages. And stopping off in New York City before leaving on a two-months-long inspection of the impact of the energy crisis on food production in Asia, Borlaug said with great sadness: "I have never been more frustrated in my life. Everything I have told Washington has been ignored. My latest analysis of the world food situation shows we are facing such a terrible crisis that 10 to 50 million people may die of starvation in the coming year. Since we have so little grain left in reserve, we are at the mercy of the weather. It boggles the mind to talk like that, but it looks as though it could happen.

"When I talk with Washington, our Department of Agriculture seems only to be worried about the return to an overproduction of cereal grain. They cannot comprehend the disaster we may be facing. Let me try to explain it in graphic terms. Picture the whole world's grain harvest as a highway—a highway of grain circling the earth at the equator. In 1971 it would have been 55 feet wide, six feet deep, and 25,000 miles long. That was the greatest harvest in history. And we ate up all of it the following year. With the world population increasing each year, we have to do more than maintain that highway of grain. We have to add 625 miles to it each year—every year, as long as the population grows. In 1972 we fell short of our total grain production goals by only 3.8 percent—which means we failed to build 950 miles of highway. To salvage the situation we nearly emptied every granary in the world. How can people talk of overproduction when our planet is in such serious condition? Don't people realize we are dealing with potential social and political chaos that can affect every system of government in the world?"

And so, incredible as it might be, the world may be facing starvation at a time when Norman Borlaug and his fellow scientists should be looking with pride upon their achievements. Thanks to their efforts, since 1948 the world's production of grain has soared: wheat has risen by 103 percent, barley by 152 percent, corn by 97 percent, sorghum by 191 percent, and rice by 88 percent. How then could this new crisis have occurred?

The blame, as Borlaug points out, must be shared by all the governments of the world. There has been little or no planning, agricultural programs have been shelved, subsidies for wheat and other cereals cut, and no comprehensive storage program implemented. Meanwhile, cities and developments around the world have sprawled over what once was productive farmland. And the richer nations have unleashed an unprecedented buying spree for meat—especially beef—the production of which uses up grain faster than anything else. In the poor countries, only about 400 pounds of grain are available to each person per year. But in countries such as America and Canada, the use of grain per person is equal to one ton a year with only about 150 pounds of this grain being consumed directly. Now, with weather experts predicting a drought in the American midwest, another failure of the monsoon rains in India, plus another drought in North Central Africa (the edge of the Sahara Desert is moving southward up to thirty miles each year), the world is dependent on the freak behavior of nature.

The interdependence of one nation upon the other can also be understood when one considers the chronology of recent events. In 1972, the Soviet Union experienced its second bad harvest in a row, proving, as one skeptic pointed out, that Marx must have been a city boy. Rather than kill off their meat-producing animals as they had done during other crop failures in the past, the Russians purchased 30 million tons of grain from the United States at bargain prices of about $1.65 a bushel. Three weeks after the purchase, the price of wheat began rising and the cost of transporting grain across the oceans increased by eight dollars a ton, because of the shortage of ships. Next, the price of wheat increased for the third world nations seeking to purchase the grain they needed to carry them through the winter (in January 1973 the price was up to $2.75 a bushel). Then, because so many ships were busy carrying grain around the world, nations such as Pakistan, which was suffering from drought and poor planning, discovered they couldn't obtain fertilizer. In the past Pakistan would order, pay for, and receive its fertilizer in ninety days. Now Pakistan found it had to order, pay for, and then wait eleven to thirteen months for delivery of the chemicals needed to grow its crops!

Meanwhile drought in the countries bordering on the Sahara continued into its sixth year and Libya lost 30 percent of its grain crop in one forty-hour period. Nearly 6 million people in six countries (of a total population of twenty-five million) suffered growing malnutrition and hunger. Other countries were desperate for grain, too. India, which gave some of its carefully built-up wheat surplus to beleaguered Bangladesh, suddenly found its own stock to be short and had to buy 2 million tons of grain at a cost of nearly $200 million.

Recently, the price of wheat has risen to the highest level in history—nearly six dollars a bushel. This means that the average price of a loaf of bread has quadrupled in two years' time. And the prospects for 1974 are for even higher prices—especially when one contemplates the problems posed by the energy crisis.

Because it takes massive amounts of energy to produce nitrogen fertilizer, farmers around the world are faced with a dilemma of gigantic proportions. Nitrogen fertilizer comes from ammonia, which is made with a hydrocarbon, usually gas or oil. It takes a ton of oil to make a ton of ammonia, which is then converted into two or three tons of fertilizer. To understand the significance of the energy crisis, consider this: in 1945, American farmers used 7 pounds of fertilizer per acre to grow corn, and today they are using 112 pounds per acre—sixteen times as much. Thus even a brief shortage of crude oil will have drastic consequences. Japan, which used to be one of the largest exporters of chemical nitrogen fertilizers, already has cut its exports by a third, if not more, a distressing development for India, Indonesia and China, three densely populated nations that depend on Japan for much of their fertilizer supplies.

In 1971, chemical fertilizer was a glut on the market, selling for $50 a ton. Recent sales in Southeast Asia have produced prices of $225 per ton. And the rising cost of fertilizer, plus its short supply, will force a nation like India to cut its use of chemical nutrients by at least a third. This will create a proportionate drop in grain production, probably a 10-million-ton drop, or a third of the massive Russian purchases.

At the last meeting of the UN Food and Agricultural Organization in December 1973 it was determined that the carryover of the world's supply of food grains was less than 100 million tons. It may sound like an ample grain supply. But it is only enough to feed the world for one month!

This means that the world is almost entirely at the mercy of the weather and its effect on the coming year's crop production. The latest FAO estimates indicate that there will be only a 3 to 4 percent increase in food production this year. But because the world's population will continue to grow at approximately the same percentage rate, because the price of food will skyrocket, and because grain consumption will increase even faster, the fact is that more and more people in the poorer nations are likely to have less food to eat.

Steps can be taken to alleviate the problem, but these measures rely on a common sense approach to the problems that Norman Borlaug has faced throughout his life. First, individual governments must make a greater commitment to agriculture. Inefficient planning must stop; inefficient government agencies must be revamped: Scandals in government agriculture programs are far too commonplace. In Indonesia recently tons of fertilizer—some of it shipped there two years ago—was discovered in a warehouse in the seaport of Tjirebon while farmers only twenty miles away were paying black market prices for the fertilizer necessary for their "miracle" rice.

Second, individual governments of the world must work for effective population control. The United Nations has designated 1974 as World Population Year so as to focus attention on birth control and family planning. But it already seems certain that most of Asia will celebrate 1974 with more births than controls. Experts from the Economic Commission for Asia and the Far East predict that by the end of this century Asia's population will double and reach

3.8 billion people—more than the total population of the world today. What is needed is a massive educational campaign. But even if such a campaign is effective population will still increase—for two to three decades if the campaign works perfectly, for a generation if the campaign works realistically, and for a century if things go wrong.

Third, the United Nations, or some other neutral party like a World Food Bank, must initiate the development of international granaries in strategic locations around the world. This has long been Borlaug's dream, but until now no one has taken his proposal seriously. If every nation contributed by storing grains in anticipation of bad growing seasons, floods, or other disasters, the world would be assured a steady supply of food. No longer would there be a six-month delay in shipping food from granaries half a world away while people starve waiting for it to arrive.

Fourth, there is no central information system that can tell us the amount of cereal grains being grown and stored in every country around the world. The Russians do not participate in the FAO. Thus, if only out of self-protection, the western nations must cooperate and keep themselves informed about forthcoming harvests. We need to know the good news and the bad news so we can plan ahead.

Fifth, there must be world-wide agreement to make fertilizer production an industrial priority. As Borlaug puts it: "Without fertilizer we're licked." The simple truth is that if the world doesn't have fertilizer, then 38 million acres of new land must be put into crop production each year just to keep pace with the population explosion. Since there isn't that much land left in the world for this, fertilizer is vital to our survival. And, at the very least, the world will have to allocate from three to four billion dollars a year for producing fertilizer, storing it, and distributing it to farmers. This expenditure is minute when compared to budgets for military hardware.

And finally, the grain-breeding programs that men like Borlaug have worked so hard to establish must be continued. The work on opaque maize and the new triticale that Borlaug has been working on has provided evidence that ounce for ounce they produce a higher level and better quality of protein than does the conventional maize or wheat. The importance of these discoveries cannot be exaggerated. Scientists are now busy testing triticale on mice. The next step will be to test it on larger animals; then it will be tested on humans in carefully controlled studies.

"In recent years," says Borlaug, "triticale has developed fantastically. But despite its potential we still haven't convinced ourselves that we have created a new grain that will place us on a par with neolithic woman, who is responsible for all the bread and cereal products that we eat today. Neolithic woman was the greatest plant biologist in history. She understood that she had to domesticate wild plants and improve them. No one knows how she accomplished these feats. Meanwhile, scientific man has never created an important new grain of any magnitude. We have worked only with what neolithic woman gave us. Triticale may be the big step forward. Its future looks bright. It may not make a loaf of bread that looks like today's loaf of bread, but it could produce bread with higher nutritional value. If it works it will be fantastic."

How, then, does Borlaug view the potential for the agricultural survival of the world?

"Half the world is looking anxiously into the skies for rain, depending on luck and hope," says Borlaug. "The other half is well fed and gives little thought to the consequences. It makes no sense to suggest that the major grain-producing nations—America, Canada, Australia—should increase their yields by 30 percent. The weak economies of the underfed nations cannot pay for the grain they will produce. This strengthens the argument for international granaries to which all nations contribute and from which all nations can borrow in difficult times.

"It is not enough to fill an empty food bowl! We need a world institution with power and impact far beyond anything the FAO can do within the limits of its international political strings. It is an indictment of the human race that we have a world bank for money but not one for food. We need an autonomous food bank that would operate on the currency of grain, making loans on judgment of needs, as men do at present with money. Many of the details need to be ironed out but it could use its income from marginal interests on grain loans to sustain the training and the work of the hunger fighters in the field. It could bolster the staggeringly low budgets given to agricultural science in all lands of the earth. It could fill out the thin ranks of functional scientists."

Now he grows emphatic, as in the old days, punching a fist into his palm, hammering home a basic philosophy:

"The world will be forced, sooner or later, to muster stronger forces to fight against hunger and famine. It is a constant, running battle—a campaign against the conflict that continues between the microbes, the diseases, the insects, the parasitic weeds, the blights and smuts. All life competes for food. Man has a place in that fight. Otherwise he will succumb to the law of evolution, which demands that you adapt and evolve, or you perish! To live, man has to eat. And somebody has to work at the science of growing food to meet the changing conditions, the changing needs, the shifty face of nature.

"Words, emotions, patterns, and shifts of public opinion will not fill empty bellies. They will not grow bread. To remove the support and backing from the men who shaped the grain plants of the future is stupidity. To place political bans on the methods, the tools, and the wherewithal of their fight against destructive attacks on human food stocks is political and affluent indulgence.

"Egypt could have been the shining star of the Green Revolution—brighter than India—with its vast water supplies and suitable climate for irrigation. But politics, hatred, and war have been stronger claimants than people's appetites.

"Other nations have suffered in other ways. The United States has banned DDT, yielding to pressures from power groups and to insubstantial allegations made without true scientific base. It has been a typical decision of convenience, made without regard for the reverberations it would have across the world in the fight for food. The weight of senior evidence was that DDT was the safest and cheapest insecticide ever produced, that it had saved a billion people from malaria, that it had wiped away pestilence such as the sky-blackening swarms of locusts which had plagued men across the world throughout history. It did not matter that there was not a shred of firm evidence that DDT has ever harmed any man, woman, or child—even those who had been in frequent contact with DDT over two and a half decades. The political decision was made in the face of a seven-month-long judicial investigation in the U.S.A.—and despite the plea made by the Indian Prime Minister, Mrs. Gandhi, that the world's underfed millions must be nourished 'even at the cost of some slight damage to the environment.'

"Must these tens of millions of human beings be allowed to starve? The ban was imposed as a political expedient, a palliative to misguided public opinion, and it has made a folly of scientific and empirical conclusions. The ban might be important to the environmental casebook, but it is an empty victory. Industrial wastes and sewage foul rivers and lakes. Our city atmospheres are poisoned. There is pollution from tobacco. These horrors are relatively untouched. Meanwhile, since there is a ban on virtually all chlorinated hydrocarbon insecticides, other chemicals will have to be used and food will be more costly. For example, the U.S. ban immediately halted the production of South American food that was to be exported for the U.S. market and contributed indirectly to forcing the price of our meat higher. There is the danger it will help create similar bans around the world before any suitable, cheap, and effective alternate pesticide can be devised. America itself might suffer from returning diseases and pests.

"Of course, America and other affluent nations have the right to make such a decision, even on emotional grounds, for if it proves damaging or dangerous, they have the power to recover rapidly. But this is not the case in the countries that come most readily to my mind."

In India, near Pantnagar and the Uttar Pradesh University, there is an area extending into the lowlands of Nepal where dwarf wheats are growing and producing seven times normal yields. Yet this area could return to a wasteland—infested with malaria-carrying mosquitoes—if it were left to nature. But, says Borlaug, "given the necessary backing and drive, given the fertilizers, the vital chemicals, and the manpower, there is no doubt at all in my mind that India can again double its grain output—wheat and rice—within the next ten to fifteen years."

Borlaug believes the same can be said for West Pakistan, provided modern technology is used to drain and reduce the growing collection of salt from the irrigated lands. He recalls that when the Pakistani agronomists gave him a testimonial dinner, one of them quoted from Robert Frost: "But I have promises to keep/And miles to go before I sleep." Both India and Pakistan have miles to go—in continuing their aggressive agricultural programs, in building schools for a more literate populace, and in curbing the monstrous population explosion. India might double its production of wheat and rice in the next fifteen years, but its population will double—to 1.2 billion people—in another twenty-seven years. Worse, new arable land that could be turned over to cultivation is severely limited.

"India will cope with these problems if it can slow down the growth of population," Borlaug insists. "If it does reduce the birth rate, then even at the present rates of food production expansion, India can feed itself—but only if there is political stability. The danger is that hunger will disrupt the stability. Anger builds into frustration, and a hungry man is not rational. There comes a time in his misery when he will kill for food—if not for himself, then for his family. India will need at least 200 million tons of grain a year in the foreseeable future. This calls for a program that must not be hampered by any kind of extremism that plagues food producers elsewhere. India's plant breeders, like all of us, will have nature hot on their heels.

"There is danger all around us. Just think, Dr. Stakman has identified something like 300 different races of rust lurking around us today—part of a vast parade of biological reactions from which one new cell can evolve and produce a crisis for man. People who claim that fertilizer stimulates algal growth, or an insecticide does this or that, that it makes soft shells and so on, should think about the enormous range of changes that took place millions of years before man was on earth. It is not the simple system that some people pretend it to be. At any time rust could break through the defenses we have erected.

"In World War II Dr. Stakman was asked by the government to look into a highly dangerous rust in Peru—a rust so devastating that it was considered to be a possible bacteriological weapon that could be used to destroy the food-producing areas of the Allies. It was so dangerous it could not be brought into North America for fear it might escape.

This same rust, race 189, is now known to exist in small areas of Italy and along the Rhine River. There are probably others we don't know about. This means that no nation can take its agronomy and its grain crop defenses for granted. Such problems, such phenomena, can be looked at only from the point of view of skilled science, not from emotional standpoints, for they are matters of life and death for millions of human beings."

Above all, Borlaug insists, the situations in India and Pakistan demand the expansion of the ranks of functional scientists to at least equal the numbers of lawyers, economists, and administrative scientists. Also, they should be given parity in rates of pay.

What is the situation in other countries?

"Move westward," Borlaug says, "and you cross an area of great variation and complexity—the Middle East and North Africa. Here the basic problems are the same. Only details vary, and important among these is the very critical shortage of trained scientists working in the fields."

From Kabul across the Afghan highlands, over the high Iranian plateau, to the Anatolian plains of Turkey it is mostly wheat country, much of it resembling the Bajío—Mexico's breadbasket—as Borlaug first knew it. "There was the same old pattern as in Mexico: the tired, worn-out soils, the poor selection of plant materials," he says. "But now it is changing, though the borderline is still close."

From these highlands down to the lowest inhabited point on the face of the earth, down to the Dead Sea Valley, in Jordan, through Syria and Turkey, the Green Revolution has taken root and the "miracle" wheats are giving new life and new hope. Turkey's record wheat harvest of 13 million tons in 1971 overtaxed the national storage facilities. Borlaug says: "It is often argued by critics of the Green Revolution that the materials and technology can work only on irrigated lands. I disagree—I disagree strongly! There has been great change in this dry region of the earth. And it is catching on across the Mediterranean basin. Go westward from Iran, Jordan, and Israel into those countries where we have CIMMYT workers in direct and indirect associations —to Egypt, Libya, Tunisia, Algeria, and Morocco. There is great change. I am hopeful that wheat crops will be doubled throughout this whole region in a matter of five years or so. But it all depends on government action and how strongly the work is backed. All these countries have vast amounts of raw materials for fertilizer production but absolutely no fertilizer industry."

He has said, many times, how it took seventeen years to build a first-class national force of agronomists in Mexico. How long must it take in the struggling poor countries, in those only recently emerging into independence?

"It is the most crippling factor we have to face—shortage of trained men. There was not a single university-trained agronomist when we started work in Afghanistan. In the whole of North Africa there was a mere handful of qualified men. Algeria, I think, had less than a dozen men in agriculture with a university background. How far could these few go without the backing of the new materials and technology and the money, equipment, and policy decisions necessary to open the new areas to food production?"

There are wheat apostles, trained in the Yaqui Valley, scattered through this whole area. Largely through their efforts the fertilizer supply has been increased, new materials have been introduced, and wheat yields have jumped well beyond all previous levels. Their work in the previous two years, says Borlaug, averted a crisis on the international front when the Russian crop disaster happened. The Mediterranean countries did not have to draw heavily on the stocks of the warehouses in France, the United States, Australia, and Canada while the Russian buying depleted those warehouses—at prices which the normally poor wheat-buying nations could not afford.

"That does not mean that all is well in these lands," says Borlaug. The food crisis threatens all agrarian nations in Africa for one reason or another. The situation across central Africa he assesses as "quite similar to the Middle East, apart from the complex patterns of disease." High-rainfall lands, moist atmospheres, and wet lowlands through central regions south of the dry Sahara where the dependence is on sorghum, corn, rice, and certain root crops—these are places where diseases flourish. "That great slab of the African continent presents a new agricultural frontier, a region bereft of technology and modern equipment, but what it needs most of all are the functional agricultural scientists who could bring change."

Throughout the African region, among the underprivileged peoples, there has been a drastic dearth of protein, the signs of which are widespread among the children. There are places in central Africa where malaria-carrying mosquitoes or the sleeping sickness-carrying tsetse flies are held in check only by DDT spraying. Without that protection, he says, the loss of manpower to work the antiquated food production projects would be enormous and crippling. Yet Borlaug is confident that a few wheat, rice, corn, and sorghum experts in the area would initiate great change, that within ten years the areas of depressed Africa would see a doubling of food yields.

As he analyzes it: "There is great controversy here too. Some conservationists argue that it is wrong to improve agriculture, because farming and animal husbandry in these areas will bring greater stress and pressure on the wildlife of Africa. My philosophy is a very different one. I will say this clearly. Hungry men will feed their children and fill their own bellies with whatever food is available. No government, no matter how much it desires to do so, will keep hungry poachers out of the wildlife refuges very long. They will fill their family pot—one way or the other. No one, no bans, will save wildlife by denying men their food. It is so simple, it should be obvious! People must be allowed to work their farms, to grow their own food, without raiding the refuges. And they must have protection against diseases —in their crops and themselves."

Pressure on wildlife; pressure on human populations. Would famine bring greater danger to wildlife than the intensive cultivation of modern agriculture? The question, he notes, is debated feverishly in affluent parts of the world, but it is not heard in the poor rural areas.

"Leap the Atlantic Ocean in imagination to the great tracts of South America's 6.8 million square miles of territory," continues Borlaug. "We are faced with the same phenomenon. Poverty springs from the roots. Food shortages mark life along the Andean highlands, in Ecuador, Peru, Bolivia, and in the heavily populated valleys of Colombia and Chile where Joe Rupert spent so many years.* They eat more corn than wheat, but the basic problems are the same as those in vast parts of Africa and the Middle East. Brazil and Argentina can be excluded from this general picture, because their situation is different in terms of unrealized potential."

During his first long journey through South America, in 1960, Borlaug became convinced of the great potential in Argentina. Untapped, not fully realized, it was one of the few lands in which he worked that actually exported food.

* Colombia had been the springboard of the Green Revolution in South America. From there Dr. Rupert extended his work to Chile. The wheats he bred from the Mexican stock made the first real inroads against rural deprivation in those lands. Dr. Rupert returned to the United States to study winter wheats for the Rockefeller Foundation and was working at Oregon State University when he died suddenly, in May 1972, from lymphatic cancer.

But the great expanse of its fertile and treeless pampas, the tracts of uncultivated plains near the Gran Chaco, the extensive capacity for irrigation schemes and soil enrichment, all teased his mind with their potential.

"I think the time is very close when expansion of output and rural production will bring changes in Argentina," he says. "It depends on whether or not there is stable government. There are some arid areas, but there are big expanses of country with good moisture from great river systems that pour billions of tons of sweet water uselessly into the South Atlantic. Production can be increased rapidly, and if fertilizer supplies are not too big a bottleneck, I would say it can be done in five years. But it all depends on the proper governmental policies."

Not only does Borlaug worry about the worldwide shortage of fertilizer; he is also concerned about the cults and faddists who fight the use of any fertilizer at all. "In using fertilizers you are refurbishing the land," Borlaug says, "giving back what it has lost through time and weather. Take phosphates: all we do is redistribute to the arable soils the phosphates that have been leached out by rains into marine deposits in the geological past. These are lifted from the sea bed and are mined to offer replacement in their original situations, on land. It is not witchcraft; it is not unnatural. But it has escaped the imagination of many who give no thought to the processes involved. The rose bush or the tomato plant cannot differentiate among its nutrient elements nitrogen and phosphorus, whether these come from decomposing organic matter, from human waste, from ammonia nitrate fertilizer, or from phosphate rock. It is all the same to the plants. Fortunately, this lack of distinction by plants gives hope for mankind."

Then there is the biggest nation in South America. Brazil's 3.28 million square miles of territory approximates the area of the continental United States. Almost half this huge country is comprised of heavily wooded regions lining the great system of the Amazon River. Navigable waterways run for more than 27,000 miles throughout the country and offer enormous supplies of sweet water for vast irrigation schemes. But despite the great access to unlimited fresh water, there are serious soil problems. In the great expanse of the *campos cerrados,* which covers roughly 500 million acres in southern Brazil, the prospect for soybeans, corn, rice, and wheat has brightened since the discovery that lime and other chemicals could be used to reduce the effect of high levels of soluble aluminum in the laterite soils.

"There are people who have argued against the use of this land for food production, saying the soil situation cannot be corrected," declares Borlaug. "But the land has now been shown to respond. Many farmers over an area of close to a million acres have treated the soils with applications of about one and a half tons of lime to an acre. New cereal grains, with shorter straws and higher tolerance to the soil conditions, will increase the already promising yields. Insects and diseases are a problem, but across this great tract of country we can overcome these problems, as we can overcome the soil difficulties. I believe that with proper controls, and adequate policy backing, Brazil can increase its food output threefold in the next four or five years. This means beating the diseases that afflict wheat and raising good summer covering crops of soybeans, corn and sorghum. But we are opposed by purists, some of whom talk about new crops such as soybeans as "exotic and strange"—plants foreign to the natural ecology. And they forget about Brazil's marvelous record with coffee, which originally came from Ethiopia. This makes nonsense of many of the arguments against opening new areas to agricultural production. In soybeans there is another promise: these plants are growing in new areas that have been treated with lime and phosphate—two perfectly natural elements. A few years ago Brazil produced about 200,000 tons of soybeans. Now it has become the third largest producer in the world, with upward of 3 million tons. The prospect is that, if the research is pursued, this output will treble again in the next three or four years. And there are a few farms producing 100 bushels of corn an acre, on plots cut out between the scrawny, miserable coarse grasses and scrubby palms that grow no more than a few feet tall, simply because of the lack of vital nutritional elements in the soils. We can do wonders there, and there are also great possibilities for livestock—if the forecasts of disaster for humanity do not hold the work back."

Brazil's government has not been idle. It has been developing an expanding program for scientifically increasing yields. But, like Argentina, the amount of backing given in proportion to the vast area of country and the population needs is utterly inadequate. And programs of incentives to the farmers, so successful in Mexico and India, have yet to be put into operation in the great lands of South America.

"We run the whole gamut of climates there," Borlaug says, "but in all those lands we have greater potential than anywhere else to produce food for the underfed countries of the world. Here is the fertility, the space, the opportunity —all waiting for the scientists to unlock their riches. The work can be done with only the slightest modification, if any, to the environment, and life can be transformed for tens of millions of people in South America. The future can be bright, provided the problems of population control can be overcome. This last will be the longest, and the most difficult, of the struggles the South Americans will face."

All these trends developing in different parts of the world will have growing impact on the more comfortable lands of North America and Europe. Britain and many parts of Europe will continue to be hurt by the price increases of formerly cheap feed grains for their cattle. The future supply of animal protein will be related directly to the worldwide scramble for fertilizers. And in the face of a probable shortage in the supply of world grain and the forthcoming steep increases in prices for wheat and corn and rice, no country in the world can sit back and remain smug.

"There can be no real and lasting comfort in any corner of the world," says Borlaug. "The dream of isolation—from the effects of the spread of hunger, poverty, and misery—will be shattered if the world does not face up to the facts. Science has given mankind a brief breathing spell—a time to catch up to his problems, to lick the population explosion, to prepare for the future. Let's not lose this precious opportunity."

What about Borlaug, what are his plans?

"I believe in the philosophy of taking the best grain we have at any given time and growing it," he replies. "We don't have the luxury of time to let young scientists tell us that this grain looks promising so let's test it. We have to grow what we have. We can test later. We need food production now, today. I will never forget what Stakman said in the first lecture of his that I heard. 'We adapt or perish.' The message was very clear, and I have remembered those words all my life. It's the reason I have fought so hard to create new food supplies. It's the reason I have to keep on fighting."

About the Author

Lennard Bickel is one of Australia's foremost scientific writers; his work has been published in many countries. In 1969 he was invited by the U.S. government to witness and record the historic moon landing of Apollo II, and in 1970 he was awarded a Commonwealth Literary Fellowship to write *Rise up to Life,* the biography of Howard Florey, who introduced penicillin to the world. *Facing Starvation* is the product of two years of intensive research, during which Bickel traveled to many countries, interviewing Dr. Borlaug and his family, friends, and colleagues. The author lives in Coronulla, Australia.